BACK TO THE BEGINNING OF A PERFECT CREATION

Discovering the Creation of Our Civilization

CHARLES PRATT

ISBN 978-1-0980-2745-2 (paperback)
ISBN 978-1-0980-2746-9 (digital)

Christian Faith Publishing, Inc.
832 Park Avenue
Meadville, PA 16335
www.christianfaithpublishing.com

Bible texts are in NKJV unless otherwise specified at a few places.

Printed in the United States of America

DEDICATION

This book is dedicated to my family. Linda is the greatest earthly treasure that my Lord Jesus has ever given me, which is my precious companion and the love of my life. I have loved her with all my heart and thanked Jesus daily for such a delightful mate and companion in ministry. Linda is beautiful on the inside like she is on the outside. I love her more each passing year. She is constantly assisting me in ministry and every mission enterprise.

Linda is the mother of our two children that have brought us multitudes of blessings and now as they are the parents of our wonderful grandchildren. Our multitalented daughter Amy and her husband Michael Spotts are parents to Ethan and Dylan. These guys love living in our old farm house in the edge of Madison County, Tennessee, in a place called Providence. Mike has taught these guys how to work with their hands and to hunt all types of critters that inhabit the woodlands in Providence. Amy has adjusted well to living with a house full of big roughed tough men. She is very talented with great computer skills and a business mind.

Jason is our missionary son who has lived most of his adult life in Southeast Asia in an unsecure nation that is extremely impoverished. While serving with these poor people since 2002, Jay has seen thousands of them give their lives to Jesus Christ. He is married to his sweetheart, Anna Goodman Pratt, who is the mother of two more of our grandchildren, Jeremiah and Moriah. When they are on this side of the world, they are at home in the Blue Ridge Mountains near Boone, NC. It is the home for Anna's family who live in Fleetwood, just outside of Boone. She has served as a missionary in India and several other nations before marrying Jay. Anna moved to Asia with

Jay shortly after their marriage in 2008. They have served with their movement partners in Jerusalem leading in new church work.

Linda and I are so blessed to have a wonderful family. Our children came to know Jesus and learned to serve Him at a young age growing up on the mission field in New Orleans, Louisiana, where they served with their parents. Charles served on the staff of Williams Blvd. Baptist Church in Kenner, where he was able to be a part of one of the fastest growing churches in the entire Southern Baptist Convention. It was in this fast growing metro church that he was able to learn some invaluable lessons about church missions and evangelism. Charles had the privilege of being mentored by the late church pastor, Dr. Buford Easley, long-time pastor and serving alongside of dozens of other wonderful staff ministers. This church served as a huge laboratory of training while receiving classroom instruction at New Orleans Baptist Theological Seminary. It was an ideal setting of classroom teachings, plus practical church experience at the same time.

I have a love for the Bible that is far superior to any other writing that has ever been written. The Bible teaches about the greatest individual in the world to me and my greatest personal hero, Jesus Christ. It also teaches that all mankind is *fearfully and wonderfully made* by the hand of God and not through some cosmic force of chance **(Ps. 139:14).** The Bible teaches that every single person is formed and fashioned by our Creator. The more that I read the Bible, the greater my love and obedience for my Creator grows. Throughout my ministry, I have tried to encourage people to read the Bible for themselves. My passion for the Creator has been fueled through a personal study each day of the Word of God. It is with a great joy in my heart that I share this message with people that have loved and encouraged me in my journey for Jesus.

The beginning of Genesis represents the basic foundation of the entire Bible. If the reader does not understand the foundational truths, then he will likely misunderstand the rest of the book. The truths that follow stand or fall upon the foundation of the first few chapters of Genesis.

This book is a personal passion of my heart from a conservative viewpoint of God's Creation of the world. Things did not just come into existence by some cosmic accident or chance, but by an Ingenious Designer-Creator that purposed every facet of His perfect creation. Many individuals have been deceived with the teachings and the lies of evolution, such as man arriving from a monkey. Some teachers have used these lies that the earth is billions of years old. However, the Bible clearly teaches that creation came to pass just over 6,000 years ago. Those years are well documented in Genesis and other Bible accounts.

Since my first degree was in science, it is my desire to show every person possible that the Bible is absolutely true and completely trustworthy. The truth is that science is attempting to catch up with biblical truth, yet it never will. That is because the Creator of this vast universe is so far advanced of all scientific discoveries that man will never be able to reach that super-human level of wisdom. God possesses unlimited understanding of creation that man will never achieve. In the beginning, every single thing made was perfect until man committed sin in Genesis 3.

My desire is to inspire all who are seeking to uncover the truth about creation to study those conservative teachings about life and how human beings arrived on earth. There is absolutely no scientific evidence of any humans living on any of the other planets. God announced early in Genesis that His focus was upon the heavens and the earth, which is taught throughout the Bible. He never mentions about man getting to another planet, but He does desire that all humans go to heaven. The provision for man to go to heaven is through the blood of Jesus, since this is the Only Way.

CONTENTS

PREFACE

It seems that no person really knows who originally made the famous statement, "If you do not know where you came from, then you will not know where you are going." Perhaps, it was Adam, Eve, or one of their descendants not far removed from the Genesis account. Every human being needs to know the origins of their life and especially the origin of human life. The truth is that men from across the ages have expressed their opinions or ideas of how things came into existence. All of those opinions could be wrong, since many are not authentic and cannot be documented with facts. Those followers of Jesus Christ have a firm foundation from the Holy Bible, which is the only reliable book with nothing but truth that has ever been written. The Lord God has authored a Book that possesses reliable, authentic, and trustworthy truth that cannot be found in any other book that has ever been written, since His is a perfect book.

This book makes an attempt to speak to the hearts of Believers that may or may not have had the opportunity to take Bible courses or seminary training in the Scriptures but have serious questions about how to explain the biblical account of creation in light of today's myriads of opposing beliefs. The primary focus is to aid church people, especially lay people in this study, since they have prayed for me and have encouraged me across my years of ministry to put my teachings into a writing about the biblical story of creation. In fact, perhaps it was a public school principal, Mr. Benny Morrison, a member of Harmony Baptist Church near Brownsville, Tennessee, that first suggested about placing my notes of creation into a booklet or a book. At that time, I was serving as the interim pastor for the church and had begun to teach on the subject of creationism, which I had developed a great passion. During that time in 1995, I was serving as the

Director of Missions for the Haywood Baptist Association in West Tennessee. No doubt, Mr. Benny saw my passion for the subject and found my teachings of great interest to him, as well as others.

Even though I am an avid student on the subject, I have never thought that there was anything new or profound that I could offer to others. That particular word of encouragement never left my mind, yet I never organized any research until I was called back into the pastorate at First Baptist Church in Kenton. I had started teaching on the subject and had been reading widely on what others had written about their research. It was during those days that I began to organize some of my studies and gave thought to putting together my teachings for some future compilation.

I don't believe that I have any special insight into God's creation. However, I do share a great interest and love of the creation story, while marveling over the Creator who made the impossible possible. My love for the subject is a growing passion in the ministry that my Lord Jesus has called me into. The events surrounding creation contains the foundation for all other Bible teachings. Since I have had the privilege of teaching and preaching in so many different places around the world, I have had countless opportunities to share with many laypersons. These opportunities have not only helped to develop a deeper love for the subject but have given much encouragement toward this writing.

Today's culture offers multiple world views that diabolically oppose the biblical view of creation. This current situation has added confusion and even chaos to many Believers, as well as unbelievers. The world definitely has their answers for how the human race ended up on planet earth; however, those views often do not coincide with the Scriptures. Perhaps, one of the multiple of Old Testament statements is found in **Zechariah 12:1**, which is so clear that God created everything. *"The Lord, who stretches out the heavens, lays the foundation of the earth, and forms the spirit of man within him."* The Lord did indeed create and make all things including mankind, which is the crown of all that God ever created. When an individual denies the Scripture references as the foundation of life, then he is denying the Creator as the Supreme God that created everything that exists.

Over the years, I have read widely about those opposing beliefs on the subject of creationism by reading literally hundreds of books, watched videos, collected articles from newspapers, magazines, heard messages, visited creation museums, plus, taught widely upon the subject. Even with all the resources available, I still feel very inadequate to write when so many scholars have written extensively upon the subject of creation. Their writings have been invaluable resources of help as I have gleaned some diamonds in the rough from those that have dug deep into the world of science. I also share a tremendous interest in science, since I hold a degree in science. This encourages those that hold an understanding of true science which follows the true scientific facts in order to discover truths concerning how things came into existence, rather than accepting an opinion of some science writer that denies any aspect of the Bible. Anyone seeking after the truth should do their investigative homework before drawing a hasty conclusion as to how creation came into existence.

Paramount to understanding the creation story, one should possess a clear understanding of who is this God of Creation. When a student of God's Word meets the Author of the Bible and learns of His nature, attributes and abilities, he will have a new perspective of God's love involved in every single step of creation. Indeed, there is a huge connection between God and mankind, but Scripture certainly declares God's tremendous superiority above man. There are some Bible declarations about God's Divine position, such as **Hosea 11:9**, *"For I am God and not man."* God is far, far superior to mankind, even though there are similarities in man's character. Man was created far superior to the animal kingdom, yet far inferior to Holy God. Understanding the character of the Creator will surely enhance man's understanding of God's eternal being and purpose for man's existence.

During part of the discussion, it may seem like a verse by verse commentary of the Scripture text, which will deal somewhat in depth with the first few chapters of Genesis. These first chapters actually explain the foundation of belief for the entirety of the Bible. The study will consider a discussion of the character and nature of God, while dealing with the facts of creation of the first man, Adam.

It is my sincere prayer that each reader will see the fore thoughts of a loving Savior in His act of creation, as Christ's redemptive work was at the heart in the creation of mankind. Since there is excellent evidence in the New Testament of God's complete redemptive plan, there are helps to clarify the full plan of "the fall of man" from Genesis chapter 3. His plan included the sacrificial blood of an innocent Lamb that would have to die a cruel death upon a cross. *"For the message of the cross is foolishness to those who are perishing, but to us who are being saved it is the power of God"* **(1 Cor. 1:18).**

This verse of Scripture has become the primary focal statement for my ministry extending around the globe. It has become a constant companion with the sustaining spiritual food in answering God's call to carry His message to the ends of the earth. In fact, our overseas ministry of **Cross Partners Ministry, Inc.** was framed upon this very verse. This verse has literally ignited a passion in my being for the Beginning of Creation and my understanding of this Divine Being that gave His life at the Cross for my personal redemption. Christ's redeeming work on the cross continues to consume my study for Him from the beginning of time. Admittedly, my passion for truth drives me deeper, so that I may be able to refute those false teachings of the devil which abound in every part of life regarding creation. Many of these attack the Bible teachings that so many hold dear to the foundation of our existence.

The devil is a consistent enemy to every minister of the Gospel and seeks to minimize our efforts in carrying out the God given mission of sharing the Divine message of the cross. Jesus referred to the devil by saying *"he is a liar and the father of it"* **(Jn. 8:44)** since his favorite strategy in the Bible is deception. He cleverly deceived Eve in the Garden from the time that he was cast from heaven to earth. Man should not be willing to place his trust in a known liar.

Before one is prepared to carry the precious Word of Truth, he needs the spiritual armor to deal with an evil minded enemy. There is no way that any man will be able to match the shrewd tactics of the prince of darkness called Lucifer, apart from the indwelling power of the Holy Spirit. After serving in dozens of different nations over the past three decades and sharing the Gospel, it has become profoundly clear that the enemy of our message has sown seeds of deception and discord in all the same places. Indeed, it is not always easy to go into a new culture and preach the truth of the Gospel. That old serpent has been actively attacking my efforts in coordinating over hundreds of mission trips in the United States and nations abroad.

The enemy tries to make the messenger and the message look foolish to people. However, nothing that he has thrown in my direction has diminished my desire and determination to continue preaching the truth as long as I have life. My prayer is that Satan will never intimidate any person from sharing the story of salvation that our Savior calls him to share. Sometimes Believers are attacked with the fear factor of failing the Lord;,however that is never from our Lord. The best way to overcome fear of failing is to keep sharing the Truth in the Word, which will build confidence in doing God's work here on earth. May this teaching become a great encouragement to every person that reads this message and never give up on serving our Lord Jesus.

May God richly bless your love and labor for Jesus Christ our precious Savior!

Charles Pratt

CHAPTER 1

INTRODUCTION TO THE STUDY OF GENESIS

There are various parts of the Holy Scripture that are misunderstood by many followers of Jesus, but the story of creation should not be since it happens to be the introduction to God, the Divine Creator, as well as His creation of all matter. Furthermore, the Genesis creation account is the key to understanding God and the entire message as the foundation of the Bible. There are hundreds of statements within the Bible, besides the Genesis creation story, that teach the truths about the matter of a Divine Creator, who brought all things into existence. One of those clear teachings is found in the book of **Hebrews 1:10–11**, which states the case, *"You, Lord, in the beginning laid the foundation of the earth, and the heavens are the work of Your hands. They will perish, but You remain."* It is worth mentioning at this point that the same Supernatural Being holds everything together that He has created. God did not take His hand off the control after creating all that He created as many have suggested. Genesis begins by emphatically stating that the Creator called Jehovah God spoke all things that exist into existence from nothing.

Far too many people of this current world have been grossly deceived about how physical things came into existence. The amazing voice commands in the beginning verses of Genesis give readers a clear and firm foundation for a belief in an incredible Creator. Many more declarations about this matchless Creator are found throughout the book of God in over half of the sixty-six books of the Bible regarding His absolute sovereignty over all the universes that He made. The Bible is very emphatic about the fact that there is only one God, Creator, and Sustainer of the universe. Because of the progress of modern science in discovering so many planets in outer space, it needs to be clearly stated that the same Creator made all of these solar systems, as well as, planet earth. In Genesis, it becomes clear that the focus of the Creator is upon the heavens and the earth. Early on everything is centered upon the creation of the planet earth with the relationship with the things in the heavens that directly relate to planet Earth.

Hundreds of profound statements are made around the center of the Bible in the books of Psalms, Proverbs, Job, and the Old Testament Prophets that declare the same Creator made all things. Consider these in **Psalms 115:15–16**, *"May you be blessed by the Lord, Who made heaven and earth. The heaven, even the heavens, are the Lord's; But the earth He has given to the children of men."* Then **Psalms 119:90**, *"You established the earth and it abides."* **Amos 4:13** declares, *"For behold, He who forms mountains, and creates the wind, Who declares to man what his thought is, and makes the morning darkness, Who treads the high places of the earth—the Lord God of hosts is His name."* It is amazing that the forty-plus human instruments that God chose to pen the individual messages came to the same conclusions that the God of creation is far superior to all things that He created. Each human writer seemingly saw an incredible, superior, magnificent and matchless Being that stood far above all things that He spoke into existence. Everything that the Creator made was complete and absolutely perfect. No part of creation, nor animal, nor man had one single flaw.

Again, **Psalms 104:5–6** states how the Lord created, *"You who laid the foundations of the earth, so that it should not be moved forever, You covered it with the deep as with a garment."* These clearly state eternal truths about an eternal Divine Creator. One of those profound truths about this Creator is found in the **Psalm 100:3**, which declares, *"Know that the Lord, He is God; It is He who has made us, and not we ourselves."* This verse alone clearly proclaims that no man was a cosmic accident, but each person was specifically planned with a purpose by a loving Savior. The Creator designed and manufactured every single thing, visible and invisible, that has ever been made. Nothing was left up to chance or happenstance, especially when it comes to the detailed attention that the Lord has given to each human being that He created.

However, there are popular notions that maybe God did create the universe, but He could have done it in a manner that blends ideas of the current science world, such as evolution, which has produced the concept of theistic evolution. There are other thoughts of theological positions that have God as the Creator; however, after all was created He took His hands off all of His work. Those that hold such a belief are referred to as Deists, and simply believe that God wound up things like an antique clock, while taking a nap until all unwinds or runs down at the end. The fact is that God had a specific plan for everything that He has created, but most specifically for human beings. The Lord teaches in **Jeremiah 29:11** that He designed a plan for each person and then He leads each person to discover that specific plan, if that person desires to follow God. *"I know the plans that I have for you," declares the Lord, "plans to prosper you and not to harm you, plans to give you hope and a future" (NIV).*

Jesus paints a totally clear picture of those plans when He declared *"My Father is always at His work to this very day, and I too am working"* **(John 5:17)**. Again, Jeremiah emphatically states the case for the continual involvement of the Creator in His creation **(Jer. 27:4–5)**. *"Thus says the Lord of hosts, the God of Israel, thus you shall say to your masters: I have made the earth, the man and the beast that are on the ground, by My great power and by My outstretched arm, and given it to whom it seemed proper to Me."* The Lord God is clearly credited with creating all things throughout the entirety of the Scriptures. Consider the truth spoken throughout the Psalms, such as *"the sea is His, for He made it: and His hands formed the dry land"* **(Ps. 95:5)**.

Another powerful declaration stated, *"You who laid the foundations of the earth"* **(Ps. 104:5)**. There are a tremendous number of similar statements found in **Psalm 124:8, Psalm 134:3, Psalm 136:5–6, Psalm 146:6.** All of the forty-plus different Bible writers seem to be in step with each other that the Creator of all things was the Lord God that spoke all things into existence in the Genesis story. It was Him that was speaking to each of these human writers as they penned the very words that became a part of the cannon of the Bible. Each writer chosen had a solid belief in God as the authentic Creator of all things, including one man. God filled up the earth with millions of creatures, but propagated the entire earth from that one man, Adam. Many of those writers made much about the human genealogy chain that reached back to that single human being.

Throughout the four Gospels and in the church letters of Paul to the Colossians, the same truth is espoused by the more than forty human authors used to pen the words given to them by Holy God, all conclude One Supreme God created it all. In Colossians, Paul states one of the strongest cases in the entire Bible for One Divine God that created all things. Paul declares that Jesus is the same God of creation in Genesis, *"For by Him, all things were created that are in heaven and that are on earth, visible and invisible, whether thrones or dominions or principalities or powers. All things were created through Him and for Him"* **(Col. 1:16).**

Then the last story in the last book of the Bible declares the same God that went to the cross for the sins of the world is the same Lord that created all things in several points in the **Revelation 10:6** states, *"Him who lives forever and ever, who created heaven and the things that are in it, the earth and the things that are in it, and the sea and the things that are in it, that there should be delay no longer."* There are hundreds of references throughout the Bible of various human writers making clear statements that the same God that went to the cross for the sins of mankind was the exact one that created everything that was created in the story in Genesis.

Thus, God's work of creation continues, since it did not cease on day 6. Each time people look at a new born baby, surely they understand that God is still busy creating living things. However, there is a sense of completeness on day 6 of all those things that were necessary for the prosperity of mankind that placed every essential into place that would lead to a continuation for all existence. That was made clear in the truth that God rested from that particular aspect of His work in the beginning. This is why Jesus declares that He is continuously working in **John 5:17**. God never takes a vacation from His activity as some have concluded, since they do not see His activity with their fleshly eyes. Without a basic understanding of who God is, a person could be led to unlimited concepts within the human mind of how all things came into existence. That is the very reason for those hundreds of worldly concepts that are being taught today.

It is always interesting to discover just what people today really believe about creation. Since 1982, the Gallup pollsters have annually asked the public those sorts of questions. According to their poll among Americans over the age of 18 released in June of 2012, revealed that 46% of those polled believe that God created human beings within the past 6,000 to 10,000 years. Gallup's findings began in 1982 with that same answer was 44%. Another question in 2012, revealed that some 32% of the responders said they believed in theistic evolution which supposedly occurred over billions of years. 15% of the adults indicated they believed that human beings were a product of random chance or evolution, while 7% had no opinion about creation **(Gallup 1).**

Later updates in 2017 from Gallup indicated approximately the same figures with only slightly different questions. For example, 80% of Americans said they believed in the supernatural virgin birth of Jesus. Another comment on the 2017 survey from the National Science Foundation study suggested that Americans should get a refresher lesson on science. That study said that three out of every four responders knew that the earth rotated around the sun and not vice versa, and a large percentage didn't know the earth's core was hot. Another question revealed a large percentage of people did not know that the father's sperm determines a baby's sex **(2).**

Because of the nature of this Genesis Study, the Gallup reports seem to speak clearly to this story. Those people saying they hold a belief in God creating all things only dropped slightly from 44% in 1982 down to 42% in 2018 on the same beliefs on the same subject. Likewise those believing in theistic evolution dropped slightly from 32% down to 31%. Perhaps, most people recognize that the current world views have changed from those of the past half century when Madeline Murray O'Hair popped up on the scene in the 1960s speaking out against prayer and Bible reading in the public school classrooms in America. Thus, many folks of the present world simply do not know which is true or maybe do not care to find the true facts. There still exists a book of facts called the Bible that could enlighten students that are willing to search for truth. Perhaps, studies on the subject of creation could be used to spark an interest for some inquiring minds. That is essentially the point of this particular book. Below is a brief report on some of the survey results revealed by Gallup Polls from their beginnings with questions regarding the creation.

Gallup Survey on Evolution, Creationism, Intelligent Design (3)

	Humans evolved with God guiding	Humans evolved, but God had no part	God created humans in present form	No Opinion
	%	%	%	%
2017	38	19	38	5
2012	32	15	46	8
2007	38	14	43	4
1982	38	9	44	9

Since Gallup Poll has been doing these surveys among Americans, some of these stats mirror previous surveys. However, one of the most interesting parts in the survey results about God having created humans in present form has decreased over the thirty-five-year history of those people polled by 6% from 1982 until 2017. Another statistic that has changed is a far larger percentage of people believing that God had no part in creation more than doubled from 9% in 1982 to 19% in the 2017 poll. This means that 10% of people over this thirty-five-year span now believe that God had no part in the creation of people, which more than doubled (from 9% to 19%) during the polling period. Perhaps, this could mean the belief of a younger generation was not in earlier polls or could it be that several people have changed their opinion? People in general, seemingly have given way to the teachings of the times and accepted that which the science world has given them is the absolute truth. One truth not on the Gallup survey is that far less people, especially among church attendees, claim they *do not* read their Bibles today and apparently have accepted the answers of science as being accurate.

One serious problem that Christians face today is all the fictional materials written in the classroom textbooks being used in public education, as well as, in many private educational systems that have dismissed the accuracy of the biblical account. It appears that the proponents of naturalistic evolution have a monopoly on the system of public education, the media and most of the culture today. It becomes difficult to attempt to tell the Bible account, when the educational system makes the Believer in Jesus appear totally uneducated or uninformed. In a Gallup survey in May 8-11 of 2014, revealed these results of preteens with some 57% believing that human beings were created by God in their present form. However, in that same survey, the stats taken by college graduates declined rapidly revealing a tragic new belief that God created humans dropped to only 27%. One huge change would certainly include the amount of exposure in high school and college professors that have intentionally left the God equation out of their teachings.

In order to debunk these teachings that are so pervasive in the modern culture, one must have the facts and prepare well to give an answer found within the Bible. Most Christians are unprepared to give the answer that seems to be missing for their unbelieving audience. The person that wants to present the truth about the origin of human life will need to do their homework beginning with the Bible as their primary textbook. The second book for the student of the Bible is the book of nature or science. Every student can profit by learning as much as they can from true science. Find as many fact books or commentaries on the subject of God's creation that support the Word of God. The Psalmist declared it well with the following statement. *"The earth is the Lord's and all its fullness, the world and those who dwell therein. For He has founded it upon the seas, and established it upon the waters"* **(Ps. 24:1–2).** Another great truth revealed in the Bible from **Psalms 100:3**, *"Know that the Lord, He is God; It is He who has made us, and not we ourselves."*

This study is of ancient history that endeavors to trace the ancestry of mankind all the way back to the beginning of the world's first human birth. Since truth is narrow, there can be only one true history of where all mankind developed. Currently there are many fabricated stories, as well as, dozens of flimsy attempts to explain mankind's foundation. This author holds a mission of revealing a single true account of man's existence, while debunking myriads of myths, which seem to be very popular in the present-day world. There is a great need to restate the case for that true version in such a politically correct culture that desires more options. Many of those possibilities will be explored in this study.

Unfortunately, the current culture is often non receptive and most often even hostile toward the idea that God created anything in this world. In fact, many folks claim they are greatly offended with such a term as a "Creator," since it does not fit the mold of political correctness. Most science textbooks have long omitted such a word, since they have widely adopted the thought that God had little or no part in the production of the universes, man, animals, vegetation, nor any design of these. The very mention of a God that had any part in this production seems absent from modern writings, leaving it up to each reader to decide the source of creation. One recent article in the **American Family Association** magazine reveals the world-view of creation is currently referred as "individualism" **(4).** Most seem content that some simple process in nature must have produced everything, which is left for each individual to determine for themselves how it happened.

If a person were to visit the Smithsonian Museum of Natural Science in Washington, DC, today, most Christians would be dismayed with the presentations in that particular museum, since God has totally been excommunicated from every part of the origin of man. From the moment that a person enters the foyer, they are introduced to their ancestors, which are the very animals that the evolutionists use in public school classrooms to indoctrinate young students. Be mindful that these places are operated by tax dollars from American taxpayers.

The common mindset of many people today is they tend to believe almost anything besides the teachings from the Bible. The attitude toward the Bible is that it opposes most of the present culture. There is a steady increase in the acceptance of homosexuality, support for abortion on demand, unwillingness to work, marriage being abandoned, clothing being abandoned, an increase in pornography, and an increase in lawlessness, to name a few **(5).** Moses continues to warn the people that he guided through the wilderness for forty years not to depart from the teachings of God. With all these admonitions and his continual warnings, the people forsook the teachings of God's Law, so a whole generation had to die in the wilderness for their disobedience to God. The same seems to apply to the present cultural, since so many well intended folks have forsaken the teachings of the Word of God during this current period.

However, the Bible teaches, "*For by Him all things were created that are in heaven and that are on earth, visible and invisible, whether thrones or dominions or principalities or powers. All things were created through Him and for Him. And He is before all things, and in Him all things consist*" **(Col. 1:16–17).** The great "I AM" of the ages stepped out of thin air and spoke all things into existence, where nothing physical had previously existed. Ultimately, these teachings are left up to each person to accept or reject the biblical facts for their beginning, since no person was there to physically witness the Genesis events, except the invisible or Spirit Being. Everything is hinged upon the invisible substance of faith. The Hebrews writer declared that *"faith is the substance of things hoped for, the evidence of things not seen*" **(Heb. 11:1).** In **Hebrews 11:3**, the writer furthered this thought by adding, *"By faith we understand that the worlds were framed by the Word of God."* The Lord's spoken Word actually created each thing He desired to create with His voice command.

Those that follow the Bible record will be offended by the present politically correct culture found in America, as well as, much of the world. The ones willing to believe the Genesis account, must believe by faith that the written record in Genesis is accurate and absolutely true. By the way, anyone that believes in any other belief present today will have to believe by faith that it happened like someone else said, such as Charles Darwin. Therefore, choose carefully in choosing a belief system that could lead a person to a wrong conclusion. It is always a good idea to check out the facts behind a person's belief system.

The most widespread and influential arguments against the veracity of the Bible is the all-too-common belief that modern science has proven evolution true, thereby discrediting the scriptural account of creation **(Huse 6).** This teaching has been widely accepted in most places around the globe, other than a few churches that have maintained the biblical inerrancy view. Most public schools have become institutions that train generations of school children in the religion of secular humanism **(7).** It is true that many people in the church have swallowed the popular view of spontaneous generation of life coming from nonlife sources such as specks of dust. Many Christians have accepted this view on the basis of its popularity, even though there is not one shred of true science in the foundation of this widely held view. Often, people simply fail to examine the theory of evolution, which holds almost no factual legs to support its farfetched premises.

If the evolutionary theory were correct claiming that nonliving sources produced everything on the planet, then life must have no purpose. On the other hand, if God created mankind, as the Bible reports declare, then human life does have a plan and a purpose. *"I know the plans that I have for you"* (Jere. 29:11) NIV Thus, the origin for life is most extremely important to the argument of where did life obtain its beginning. This study is all about arming followers of Jesus Christ with information that will enable each believer to stand firm against the falsehoods of the day. If life is not embedded in the work of Jesus, then He was a fool for dying on the cross for the sins of Adam, as well as, all mankind.

There is a thought that might connect with some readers with the Bible message. "If there is no Creator behind all of creation, who has set the absolutes, then why are people still following these Christian rules about marriage, sex, truth and ethics? Why not just do whatever feels good without experiencing any consequences?" **(8)** Perhaps, people know deep down within their being that there are moral absolutes of right and wrong, especially, if they grew up with those conservative teachings before things became so politically correct. Actually, the Bible teaches that it is the Word of God that changes the human heart, when a person is informed or reads that Word from God. If this is true, then the children of God that are true Believers are responsible for taking the Word of God to those that have not received Jesus into their lives. Actually, this is a commandment within the Word of God in the Great Commission in **Matthew 28:19:** *"Go therefore and make disciples of all the nations, baptizing them in the name of the Father and of the Son and of the Holy Spirit."*

The answer is clear that America, unlike most nations, was founded upon Judeo-Christian convictions of right and wrong from the teachings of the Bible. There is a God that has given order and organization to everything in creation. These teachings are instilled in most people today through their ancestors that everyone is accountable for their actions with consequences that follow every decision. Those teachings began when the Creator taught those first humans, Adam and Eve the rules to live their lives. However, the Bible says, *"The fool has said in his heart, there is no God"* **(Ps. 14:1).** It is no wonder that the people of today's world are in serious trouble, since they believe there is no God, thus they are free to live their lives as they please, without any consequences.

Mankind denied God from early accounts in the Genesis story, especially by the time of the Great Flood in Genesis. The Jewish people that descended from Abraham were different, since they were a chosen people that the Lord God instructed to teach their heritage to every generation. Every Jewish family would teach their offspring their heritage as God directed in **Deuteronomy 6:7–9**. The chosen people of God talked often to their children about their history by writing these teachings down on the doorposts of their houses, on frontlets between their eyes and gates to their homes. Their history was passed on orally from one generation to another in that manner, so that nothing of their history was lost. The children were taught to follow God's laws, since He was the Creator of all things. These accounts were rehearsed until every offspring knew their heritage and knew it well enough to repeat it.

Perhaps, a bigger question for consideration about creation is what a person believes about the Bible and its accuracy. For example, there are many of the church today that believe that the first eleven chapters of Genesis are an allegory, rather than a factual account of the beginning of all things. The serious student of the Bible needs to settle in their heart what they will truly believe regarding the authenticity of the Word of God. It matters greatly whether a person takes the Bible in a casual fashion or is serious about studying the Bible for their own needs. Each person must decide if they truly believe that every single word in the Bible is true and trustworthy for them.

The Bible is abundantly clear in multitudes of places such as **Jeremiah 10:16**, when he declares, "*For He (God) is the Maker of all things.*" Genesis is the seedpod of the Bible or the embryo of every major doctrine that is found in the first book of the Scripture: the origin of the universe, the earth, Homo sapiens, right man, right woman, dispensations, sin, death, redemption, divine institutions, laws of establishment, nations, civilizations, and the nation of Israel **(Thieme, 9)**.

If a person does not view the first eleven chapters of the Bible as authentic and exactly accurate as God called the worlds into existence, then that person may likely decide that other parts of Scripture are untrustworthy. Each person should consider the Bible and how they will accept it. The Bible is not a science book, but it certainly contains a lot of science. People of the Old Testament and until the time that Columbus sailed across the Atlantic believed that the earth was flat, rather than a sphere. Yet the Old Testament clearly declares three times that God sits or walks on the circle of the earth. "*It is He who sits above the circle of the earth*" **(Isa. 40:22)** then in **Proverbs 8:27**, the Bible declares that "*God drew a circle on the face of the deep.*" Then in **Job 22:14**, "*and He walks above the circle of heaven.*" Scientists believed that the earth was flat until about 1492 AD. The truth is that science is attempting to catch up with the teachings of the Bible. Thus, the Bible is not a science book, nor is it a history book. However, the Bible contains a lot of the history of civilization from the very beginning to the end. In fact, the word Genesis means generation, beginning, source or origin. This essentially suggests that the Genesis account is all about the origin of mankind and especially human life.

Many theologians have agreed that the book of Job may be the oldest of the biblical writings and preceded the writing of Genesis. Here is a thought from Job as he endured his days of suffering. Job said, "*God alone spread out the heavens, and treads on the waves of the sea; He made the Bear, Orion and the Pleiades, and the chambers of the south; He does great things past finding out, yes wonders without number*" **(Job 9:8–10).** Indeed, the Creator can show the vilest person His supreme existence through His creation. Man can manipulate others, but God is able to change man's heart, since He knows the very thoughts of every man **(Psalms 139:2).**

If God could speak the universes into existence with a voice command and is indeed the Creator, then He is the Owner of all that exists, including man, who must be only a steward that is accountable to the Owner for all of his resources **(10).** Then man must be in the debt to his Owner of all that he possesses. A functional Bible principle, which exists is that man will have to answer to God one day for every wrong that he has committed. The Bible declares that all men have an appointment with death, but after this comes the judgment of God **(Heb. 9:27).** Every man will die one day, then immediately face the Creator to be judged for the life that he has lived.

Sometimes when a writer mentions that a Creator was the source of all existence, generally the publisher of the document will be contacted and asked to edit out such a suggestion of God's design. Threats could be lodged against the publishing company declaring their books will not be welcomed on the store shelves with such an outrageous narrow claim or statements. Some will even claim that the publisher has violated "the rule of science" by using the term Creator. Those that make such charges are often angered with words such as "supernatural" when referring to the system of "natural science." Natural carries with it the idea that things merely came into existence by nature and chance. Currently, the world teaches that nature has guided the entire process over a course of billions of years. Statements that do not fit this scenario often suggest that the writer is simply uneducated or needs enlightenment. This condition has been fostered by situation ethics and more so by the political culture that exists in today's world. This modern-day situation demands that no speech or writing should offend any person, yet it commonly offends the majority of people that follow the teachings of the Bible.

The Bible is very specific about the "Source of Life," whether regarding animals, plants or human life. So the student of the Bible and those that follow the Creator view must be willing to face the challenges of a politically correct society that opposes the God of Creation and His absolutes. Those people that teach tolerance are almost always intolerant of Christianity **(11).** Those that follow the biblical teachings need to prepare themselves with some answers that will not further anger their audience, but will challenge their position on creation. One may want to ask questions such as, "Well, how do you think the highly sophisticated structure of the human body came into existence?" or "what do you believe that holds all of the planets into their rotation in the atmosphere around the sun?" Sometimes the nonbeliever in the Creator concept will respond to the question idea rather than statements of fact. True Believers should never reply to the skeptic with anger towards that person, but always give kind replies. Likely, the person is without Jesus in their life or has not yet heard the truth explained from the Word of God. Often, they have readily accepted a teaching that they may have heard or read from a book by an author that opposes the Bible.

Remember that most writings today merely assume that everyone has accepted that creation came through an unguided natural process of science, the Darwinian process of evolution or some of the other processes that will be discussed in chapter 13 dealing with the common theories of creation being taught today. A student of the Bible account on Creation would be wise to investigate the common beliefs of his day, so that he may know how to reply to the questions that others possess. However, there is a caution for any person studying other beliefs. They should have a sound understanding of the Bible before digging into these pagan beliefs. The enemy to every Believer is the same enemy that deceived Eve in the Garden and he continues to deceive people today in like manner. His tactics worked in the Garden, thus he continues to use those same approaches throughout the Bible by playing upon what people may not have learned about God's Word. Every person needs a certain foundation for their belief from the Bible before pursuing the belief of the enemy of the Bible.

The Bible, itself teaches that it is a perfect book without any flaws or errors. The term that sometimes used in the Scriptures is called infallible or inerrant, which means perfect or without any degree of error. Psalm **119:140** states that case when it says, *"your Word is very pure (refined) and therefore your servant loves it." "The entirety of Your Word is truth"* **(Ps. 119:160).** *"Forever, O Lord, Your Word is settled in heaven"* **(Ps. 119:89).** The Psalmist further declares that, *"Your Word is a lamp to my feet and a light to my path"* **(Ps. 119:105).** Since God, the Creator is perfect, thus He will always speak perfect truth. His Word is full of truth, so man may always trust the Word of God. Thus, Genesis is an unembellished, chronological record of this world, before there were historians to record the events making the Bible of man's beginning **(12).**

In the New Testament, Paul stated it to the young preacher Timothy, *"All Scripture is given by inspiration of God, and is profitable for doctrine, for reproof, for correction, for instruction in righteousness, that the man of God may be complete, thoroughly equipped for every good work"* **(2 Tim. 3:16–17).** No Believer will be prepared to engage those without Jesus until they are filled with the Word of Truth, since it takes the Bible instruction to make one "complete." Therefore, no person is adequately prepared to face the devil, nor those worldly minded, until he has completed his daily preparation in the Word of God. No wonder so many fail, because they are incomplete apart from the Truth of God's Holy Word.

Genesis derives its name from the original Hebrew language of the Old Testament, which means "Generation," "Origin" or "Source" in English. That word transliterated is *"Bereshith"* that is "In beginning." This writing reveals the "source" or "origin" of all creation and how supernatural things came into existence, rather than by a natural course of chance. Commonsense reveals that something or someone far superior to humanity constructed all that man can see and experience in the universes. Again, the chances of all the complexities of life accidently falling into place seem like a fairy tale.

In fact, man has already discovered in early centuries that certain laws exist that are commonly referred to as the Laws of Nature, the Laws of Physics, but more correctly the Laws of God. Scientists have long ago agreed that these laws operate on earth, such as the Law of Gravity, yet scientists have no idea of how these laws became operative, nor came to exist. Proponents of natural selection have no answer for the beginning of anything. Many people blindly follow the natural progression or teachings of Charles Darwin, who was one of the most popular advocates for an upward evolution for all things from some microscopic speck of lifeless dust over a period of millions of years. Darwin, who was trained in theology at Cambridge University, offered the modern world an unproven answer for the beginning of all life in his poplar book entitled ***On the Origin of the Species***," which he published in 1859.

Perhaps, no other book has had such a negative impact upon the thinking of man to lead mankind away from God's teachings on creation as this book by Darwin. Highly educated people use these teachings as absolute fact. Darwin contended that light would be thrown on the origin of man. Sure enough about a century after Darwin published his book, the light of the science of evolutionary biology which Darwin founded, man is seen not just as a part of nature, but as a very peculiar and indeed a unique part **(13)**. Prior to Darwin hypothesis, the world of naturalists had believed that the species were immutable productions and had been separately created **(14)**. Charles Darwin's brand of science has had a powerful impact upon the world of science that so many people have refused to question his beliefs.

Darwin on a chapter of "Natural Selection," presupposes that everything evolved from something else over long periods of time of divergence from one order to another order, whereby there are no concrete facts that such things happened. He uses one story about grass sod that produces many other species of grass over years and years. Darwin claims, "He found that a piece of turf three feet by four feet in size, which had been exposed for many years to exactly the same conditions, supported twenty species of plants and these belonged to eighteen genera species of plants, and eight orders, which shows how much these plants differed from each other" **(15)**. Darwin was actually telling his audience that a species or characteristic not only changed in plant life over a long period of time, but that the order of the plant life changed, too. While true science may agree with mutations that do take place in plant life, but no plant grows into another species or order of plant life. There is no science to support such a happening. *Webster's Dictionary* defines species and an order as another "kind" of plant like the grass becoming an upright tree or turns itself into a shrub over a long period of time. It is true that grass can and does take on different features but continues to be in its own species of a process of changes called mutation or "micro evolution." This is referred to in the science world as "micro evolution or small changes in appearance, but not "macro evolution," which would be major changes as grass becoming a tree.

Later, Darwin shows this same phenomenon happening in animal life. There is zero evidence to support a species change of major proportions or an order change such as a cat changing into a dog over long periods of time or a frog becoming a dog. The nature of each plant species was given by the Master Designer placed all things in an order. The Creator of this vast universe is a God of order as He spoke all things into existence for the benefit of mankind. The Lord shows forth His order in everything that He designed and purposed. Many of the species of plants and animals have gone away into extinction. Darwin got it wrong with evolution and now his teachings are likely far more common than God's story in His Book that He wrote called the Holy Bible.

Man, who is "made in the image of God," seeks true answers for life that brings satisfaction to his inner being, since a deep sense of God is within man's being. There are always some people that offer an answer or solution to man's question. However, when a person begins to seriously look around or looks upward, God is able to speak to that inner being of mankind. God shows seeking man plenty of evidence through His beautiful creation that surrounds each person. That person that believes in God as Creator might ask a question of how God did all that was essential for the creation. From the Bible, man is taught that before the "Beginning" the universe was not, and that "the worlds were framed by the Word of God," according to **Hebrews 11:3 (16).**

Again, the problems abound when man's imagination runs with the popular ideas in today's culture. One such idea that must be dealt with is the fact that the Creator created individual species of creatures or did all creatures come from the same ancestor? Darwin produced the thought that man was the product of some intermediate ape man. There is zero evidence of this ever happening, since there are no fossil records of the beast which kept changing from an ape to that of man. Billions of fossil records are buried within the earth's crust around the globe, yet not one single fossil exists of those so-called millions of years while the transition was taking place between these two creatures. Why has no geologist or scientist not discovered any of the fossils from these so-called intermediate creatures?

A simple study of the skull of a man and that of the ape give different conclusions, suggesting that there is no connection. The human skull is easily distinguished from all living species, though there are, of course, similarities **(17).** The large brain size in humans is very evident from the small brain of the ape. The front of the skull of the human is always vertical, indicating that he is an erect creature, whereas the skull of the ape is smaller with sloped head from the top to the chin **(18).** None of the skull shapes have ever changed over the many years in any of the fossils that have been studied.

In conclusion, it can safely be stated that there are two absolute certainties regarding God and the animal kingdom. For those that are still seeking absolute evidence that God exists, there is no absolute scientific proof to satisfy the inquiring human mind that God really exists. The second absolute is the belief about monkeys evolving into humans. Since the beginning of the human race, there is a mountain of evidence that humans only produce human beings. These facts have been witnessed millions of times, but never has there been a single observance that a human being has ever produced a non-human being. These facts keep mounting every single day as more human births are being celebrated around the world. God's Word has never, nor will ever be proven to be untruthful. After the studies and research all rise to the surface, God will still be the Divine Creator of all things that exist. God is the only one that can create something out of nothing.

Chapter 1

Introduction to the Study of Genesis

1 **www.Gallup.com, Gallup Poll Survey,** Views of Origin of Human Beings, May Survey Study, 6-1-12.
2 Live Science, ATechMediaNetworkCompany, **Science News**, March 8, 2017.
3 **www.Gallup.com**, **Gallup Poll Survey**
4 **AFA Journal,** American Family Association, US Evangelicals Confused About Theology, June 2017 issue, pg. 8.
5 Ham, Ken, **The Lie: Evolution**, Genesis, the Key to Defending Your Faith, Master Books, Inc., 23rd printing, Green Forest, Arkansas: 2002, pg. 17.
6 Huse, Scott M., **The Collapse of Evolution**, 3rd Edition, Baker House, Third Edition, Grand Rapids, Michigan: 1997, pg. 17.
7 Ham, **The Lie: Evolution,** pg. 71.
8 Ham, Ken & Taylor, Paul, **The Genesis Solution**, Baker Book House, Grand Rapids, Michigan: 2000, pg. 45.
9 Thieme, R. B., Jr**, Creation, Chaos, and the Restoration,** R. B. Thieme Bible Ministries 1995, pg. 1.
10 Morris, Henry M. III, ***Book of Beginnings***, Vol. 1, Creation, The Fall, and the First Age, Institute of Creation Research, Dallas, Texas: 2012, pg. 13.
11 Ham, ***The Genesis Solution***, pg. 49.
12 Morris, ***Book of Beginnings***, pg. 9.
13 Darwin, Charles, ***The Origin of Species***, New American Library, a Division of Penguin Books, Ltd., 80 Strand, London, England: 1958, pg. 18.
14 ***Ibid.***, pg. 19.
15 ***Ibid.***, pg. 107.
16 Spence, H. D. M. and Exell, Joseph S., Editors, ***The Pulpit Commentary***, Vol. 1, Hendrickson, Publishers, Peabody, Massachusetts: pg. 7.

17 Menton, David, ***Apemen***, "Separating Fact from Fiction," Answers in Genesis, Petersburg, KY., 2010, pg. 13.
18 ***Ibid***, pg. 14.

CHAPTER 2

TITLE AND AUTHORSHIP

The book of Genesis focuses upon the origin of the universe, the earth, and mankind in chapters 1–11, while chapters 12–50 deal with beginnings of the Hebrew nation of people. It is only appropriate that the first section of the very first book deal with the beginning of ancient civilization and some of life's toughest, yet most basic questions about life. For this study on creation, the first eight chapters of the Bible will be discussed to help answer some of those significant questions that are being asked about creation and life. The study will include the teachings surrounding the Great Flood, since this is the judgment of Holy God upon those that reject Him and His commandments. The Flood of Noah's day is such a critical event that literally changed the whole world's environment and has become such a highly controversial issue throughout the centuries. It is so connected to so many questions surrounding all that happened to the creation that it must be a big part of the discussions that follow.

One of the biggest questions about the creation account is whether the first chapters of the Bible are to be viewed as literal happenings or a figurative message. The Bible answers those questions by giving real names and ages in the genealogical list in Genesis 4, 5, and 10. It simply does not stand to reason that such names, places and ages would all be given if they were symbolic or representative of the real characters. In any literature, the literal always precedes any symbolic language, therefore the creation account cannot be symbolic or metaphoric. The biblical account simply informs the reader how the events of creation actually occurred, while revealing how a loving Creator brought all things into existence.

All evidence points to a literal account of civilization with no hidden message in the writing, plus the fact that later Bible writers refer to these as real people. Jesus spoke of many of these as real people in the gospel accounts. He referred to Moses as the human writer of books of Law, when Jesus discussed the reason that God gave a writ for divorce. In **Mark 10:1–12**, Jesus was in Judea teaching a multitude when a Pharisee asked a question regarding the Law of God. Jesus simply asked the Pharisee, *"What did Moses* (the Lawgiver) *command?"* They answered that Moses in the Law permitted a man to write a certificate of divorce. This is one of the clear points where Jesus and the Jewish leaders both agreed that Moses was the one that wrote the Peneteuch (first five books of the Bible or the Books of Law). Jesus only taught truth and never taught His followers fables and allegories. He commonly taught in parables, which are life stories that many of followers personally knew, which contained a heavenly truth that Jesus wanted His audience to clearly understand.

The writer of the book of Hebrews also wrote that these were real characters who acted in faith to believe and to follow the Lord God. For example, the writer's list mentions four of these early Bible characters in the "Hall of Faith" chapter of Hebrews 11. He names Adam, Abel, Enoch, and Noah and commends each for their faith in God. This passage of Scripture alone would deny the belief for an allegory or the basis for a fictional story of any sort. Thus, the first eleven chapters of Genesis is an actual account of a literal message of how man and early civilization came into existence.

The authorship of the first book is often debated, while many Bible scholars suggest there were multiple authors, others contend that one author simply compiled writings of several people, some say that the author or authors used different sources for the content of the book, while some contend that Moses was the human author that God gave a Divine revelation. Genesis actually contains no statement as to who wrote the story. Most of the titles of the sixty-six books of the Bible were likely given at a later date than the writings, and most titles were derived from the message or the writer by later Bible scholars. In Genesis, the title comes from the very first word in the original Hebrew language of the book *(in beginning)*. Martin Luther made a strong statement when he substituted in his German Bible the title "The First Book of Moses," a designation requiring no further commentary **(1)**.

There are many critics that contend for the multiple authorship theory and that they used many different sources of historical writings and legends that were available to them. Those source materials are often referred to as the J, E, P, and D sources. Some theologians claim that these were actually different persons, while others believe that these four sources were schools of thought. Most of these source materials were developed from the different names the writer uses for God. For example, in the first creation story in **Genesis 1:1–2:3**, the author uses the reference of "God" (*Elohim is the Hebrew word*) thirty-five times. In the second creation story of **Genesis 2:4–25**, the title of the Divine Creator is "Lord God" (Jehovah or Yahweh is the Hebrew word for Lord), which is used eleven times. Then in the fall of man account in chapter 3, the title begins with the title "Lord God", but quickly changes to "God," while the last two thirds of the chapter, the writer returns to the term "Lord God" again.

The abovementioned J theory refers to the Jehovah or Yahweh sources, the E theory refers to the Elohim source, the P theory refers to the priestly sources, while the D theory is a reference to the Deuteronomist source. The different references to the Divine title continues to change throughout the writings; however, the Hebrew people very commonly interchanged the titles. Several studies have been done on the various names for God used in the Bible and most have concluded that there are over 250 different titles used, while others count over 500 names.

Even though all these different names were used and that several sources were utilized, one writer could have easily completed the writing of Genesis and the five Books of Law. If that is the case, then Moses would definitely be the leading candidate for the work. Whoever the human author was may not be that significant as long as one believes and understands that God authored the entire Bible.

He literally breathed into every human writer the very messages, phrases, and even the very words to share with His people. The Bible says, *"All Scripture is given by inspiration of God, and is profitable for doctrine, for reproof, for correction, for instruction in righteousness"* **(2 Tim. 3:16).**

The next question that people often raise about the validity of the writing is, how did Moses obtain the knowledge of the events of creation, since he did not live until about 2,500 years after the fact? There are several different possibilities. It is possible that Moses received the whole account by direct revelation from God, or he could have received it from ancestors who received the story by revelation and accurately gave it to him, or it could have come as oral tradition from his forefathers who had rehearsed every detail. Most parents attempt to teach their offspring about their heritage, especially the Jews, since they greatly valued their historical roots. The Jewish people truly believed that they were "God's chosen people," so they cherished their heritage.

First of all, the links in the chain of persons from Adam to Abraham were very few because of man's longevity of life, and Abraham's time was already one of intense literary activity; secondly, Godly men who perpetuated this tradition would have employed extreme care to preserve it correctly in all its parts; thirdly, the memory of men who trusted more to memory than to written records is known to have been unusually retentive **(2).** Whichever was the case, the Divine Spirit of God was watching over the accuracy of events while utilizing those writers that He chose, since the story contains the full and absolute truth.

Perhaps, one of the strongest pieces of evidence for the Moses authorship is found in **Joshua 8:31–35**. At the triumphant celebration of the defeat of Ai by the army of Israel, Joshua led the people in worship where "the Book of the Law of Moses" was read at the foot of Mount Gerizim and Mount Ebal. During that time of worship, Joshua read all the words of the Law of Moses to the children of Israel. The first five books of the Bible have been referred to as the "Books of Law," far back in Jewish history, which would clearly indicate Moses had written all of these works prior to the crossing of Jordan and likely after he left Egypt the second time with Israel.

For the sake of any further discourse of all the many, many opinions, this author will gladly accept Moses as the instrument that God selected to pen the Genesis account, as well as, the other four Books of Law (Exodus, Leviticus, Numbers, and Deuteronomy). Moses had the best of training and education of any Hebrew that is mentioned in the early part of civilization. He further spent much time alone with God in secluded areas while praying and fasting. The Bible also declares that Moses was a very humble and meek man that was usable to the Lord. Moses seems to have the logical edge among all the Hebrews mentioned as possible authors of the era.

Chapter 2

Title and Authorship

1 Barnes, Albert, ***Barnes Notes of the Old and New Testaments,*** Exposition of Genesis, Vol. 1, Baker Book House, Grand Rapids, Michigan: 1981, pg. 5.
2 **Ibid.**, pg. 36.

CHAPTER 3

FREQUENTLY ASKED QUESTIONS ABOUT CREATION?

One popular method of teaching truth is to ask questions, which is a favorite of many teachers, as well as, this author. It provokes the student to think about the subject being discussed. There are a few dozen or even hundreds of questions commonly raised on the subject of creation, particularly by those who hold a view that does not agree with the biblical teachings from Genesis. Here are a few of some very popular questions that will invariably surface in a discussion or teaching time.

These are not in any particular order of the questions that this teacher has been asked over the years when teaching on the subject, but some of these have continued to pop up very often. Some of the questions that have been asked more often than others in the list were purposefully not included, since they will be answered in the text of the first four chapters of Genesis. Invariably, people seem to ask questions such as: "Where did Cain get his wife?" or "What was the mark God placed on Cain?" Questions like these would certainly be included in most of the **Top Ten Asked Questions**. However, at this point most will receive a brief treatment, since those with Cain will receive a more serious study in later discussions, especially in Genesis 4 (chapter 10, entitled "Raising a Little Cain"). The top ten questions will certainly be different at every meeting house, so these are some of those most common topics for discussion during the teachings of this author.

Some Commonly Asked Questions

1. Did dinosaurs exist? Did they live with man? What happened to dinosaurs?
2. Is the earth billions of years old or was it created more recently?
3. Did men really live for hundreds of years in the Old Testament times?
4. Is there any evidence for a Global Flood? Was there ever an Ice Age?
5. Did all the animals get on the ark?
6. Did God create the world in 7 literal 24 hour days?
7. Who created God?
8. Was there another world before creation?
9. What was the earth like before the Flood?
10. Do angels really exist? What is the purpose of an angel?

Dinosaurs—Yes or No?

These are brief responses, since most of these have more complete answers appearing in other chapters in this writing. The term *dinosaur* was not coined until 1841 by Richard Owen for a reference for enormous size creatures that have always been sources of conservation, apparently since the beginning of life on earth. One of the oldest writings in the Bible, if not the oldest according to many theologians is the book of Job. Remember that Job lived a long time after Adam and Eve. In fact, most early theologians believed that he lived in a period after the Great Flood of Noah's day. If this is true, then this should become a very important discussion with those Believers in Jesus to be able to refute those that claim in another world before this present world. The Bible dating system is very clear that the flood occurred 1,656 years after creation, which happened about 2,344 years before Jesus walked upon the earth. This would mean that the earth is approximately 6,000 years old and that these giant dinosaurs lived with man and not millions of years ago.

Early in the tribulations of Job, he expressed how that his desire was to erase the very announcement of his birth that his mother would have been barren and there would not have been any joy surrounding his birth, because of a good life that turned into great grief. During that time of tremendous sorrow turned into a great learning experience for Job and everyone that reads his story. It was during those sorrows that the God of creation asked this man some tremendous questions concerning his knowledge of creation. Had Job's hardships not befallen him, then maybe there might be more questions than answers about these huge creatures that roamed the earth during his lifetime.

Bible characters like Job made comments like rousing up a tremendous size creature from the earth called Leviathan **(Job 3:6–9). Isaiah (27:1)** also knew of these enormous-size creatures, when he declared that God "*will punish Leviathan the fleeing serpent, Leviathan that twisted serpent; and He will slay the reptile that is in the sea.*" This monster of the sea was not just a legend to early mankind, but was a very real creature that was feared by men that only God could control. Remember that Isaiah's prophetic ministry was carried out about seven hundred years before the coming of Jesus Christ. If a person believes the biblical accounts of Isaiah's life, this would have been a little less than three thousand years ago and not billions of years as some writers have suggested.

This creature is later mentioned when God begins to question Job **(41:1–34)** that he could sneeze out flames of fire and smoke like some huge dragon that no man could tame. Leviathan could not be taken by any man nor beast as God asks Job, what would Job do with such an enormous creature that the Lord God had created. Job knew of this monster, as well as, another one that the Lord quizzed Job about called Behemoth which had a tail the size of a giant cedar tree **(Job 40:15–24).** These huge creatures were present with man as the Psalmist said **(104:26)** "*in the oceans, where the ships sail about, there is that Leviathan, which God made to play there.*" This could have been King David, since it appears in the midst of some Psalms credited to him or a writer during that same period of time.

It is very possible that all of these Scriptures were written after the Great Flood of Noah's day, meaning that some of them survived the Flood of Noah. In fact, if one believes what God stated then certainly a pair were on board the ark. In **Genesis 6:19–20**, the Lord promised to preserve all living flesh of at least two of every species of creatures would be inside the ark. God said to Moses, "of the earth every kind, two of every kind will come to you to keep them alive **(v. 20).** This would place some of these enormous creatures that God created on day 6, living in the recent past only a few thousand years ago and not millions of years ago. Many fossil records have proven that such huge creatures lived a few thousand years ago and not billions of years. A student of the Bible needs to decide whether he will believe and trust in the Word of God or in the writings of some so-called experts of science. One of these has to be wrong, so that decision will rest upon the faith of man to decide and who each person will trust to guide them. Most of those that contend these enormous creatures died out billions or millions of years ago have made their argument with a dating system that has been proved seriously flawed with multiple errors.

From the Bible records, these huge creatures lived during the early days of creation along with man inhabiting the same earth. God created every single thing that has been created, and thus, He created these creatures on day 6, just prior to creating the first man. Footprints of very large creatures as these described in the Bible exist, as well as, fossils from such creatures all over the surface of the earth. The next question is, "What happened to them?" Most of them likely died off in the worldwide Flood. The entire climate changed during the Great Flood in the day of Noah when the very first rain fell upon the earth. Not only, did the water vapor canopy (windows of heaven) surrounding the atmosphere of the earth break apart, but according to Genesis the great deep was ruptured. The entire crust of the earth broke apart at hundreds of points around the earth, allowing water to surge from the belly of the earth literally spewing upon the earth at the same time the windows of heaven broke open.

Some of these enormous-size animals were found buried in ice at the poplar caps whole with green grass in their mouths, which is a clear indication that they died suddenly from some cataclysmic change in all of their surroundings. There are many people today that believe that an asteroid flew through the sky and hit the earth killing all the dinosaurs about 66 million to 100 million years ago. This theory is full of holes such as where is the crater located that crashed into the surface of the earth and how did that one asteroid kill off all the dinosaurs on the opposite side of the planet by bombarding one single place.

There are believers in the asteroid theory that suggest the dinosaurs had gathered in a convention of dinosaurs from all around the earth to some place in modern day southern Mexico when this crash into the earth occurred. Thus, all the dinosaurs perished at the same time. One big problem after another exists with this theory. The remains of these animals have been found on the surface of the earth all over with fossil decay of their skeletal system that suggest a time frame of about the period of the great Flood, which happened about 4,500 years ago rather than 66 to 100 million years ago.

The state of the fossil records does not support this long length of millions of year time frame. Flexible animal tissue still persists inside many fossilized dinosaur bones. Blood vessels, hemoglobin proteins, and whole bone cells could never last one million years, let alone 66 million **(1).** It is safe to suggest from scientific research evidence available today that the Great Flood buried many of these animals under a deluge of mud rather quicker during the day of Noah about 4,500 years ago.

Prior to the Great Flood, the earth was likely a pleasant paradise with temperatures all around the globe at about 72 degrees, including the north and south poles. With such huge changes in the atmosphere after the flood, it was much more difficult for these huge animals to survive. The initial creation was very rich in oxygen with the water vapor canopy, but all that changed drastically during the flood. The larger animals needed far more oxygen than other animals. Because of so many changes in the atmosphere, some have suggested that these animals may have become much smaller over the next few centuries than prior to the flood. In fact, most scientists believe that the Ice Age began following the flood, which produced an unsuitable situation for many different kind of animals.

Remember that God made a perfect universe and all that He created was perfect, including man. After man committed sin and rebelled against God, then things began to go downhill. Man became increasingly more sinful with every passing generation, thus bringing on further judgments of God. That disobedience brought forth people that refused to honor and worship the Creator, so He sent forth a worldwide flood to destroy all mankind, except a righteous man named Noah and his family. The righteous judgment of a Holy God fell against unholy sinful man upon all the world, which dramatically changed the whole landscape of the planet.

According to the Bible, God broke open the windows in Heaven with such a fierce storm or deluge of water which no one could withstand and live through. At the same time, the Lord broke open the belly of the earth permitting waters to spew out with such force that almost nothing would survive outside of the ark. This drastic change in the earth, the temperatures and the force of the devastation of God's judgment upon the earth made it impossible for man or beast to survive, other than what was safe onboard the ark. The waters from inside the belly of the earth were likely very hot that helped to immediately change the entire environment.

Fossil records prove that these huge creatures lived during those early days when man walked upon the earth. These huge creatures did not live in a prehistoric period of time before man as some scientists and others have claimed. Again, a student needs to believe the Bible account or choose a false narrative that omits a Divine Creator. There is more on this subject in chapter 7 on day 6 with the animals that God created just before He created man. Additionally, the expanded discussion in chapter 13 of the theory of evolution contains further information.

How Old Is the Earth?

The biblical statements will be given regarding the "Age of the Earth" in chapter 12 of this discussion. The Bible contends for a much younger earth with the period of man only slightly over six thousand years from the biblical account of creation to the present day. The non-Bible accounts have been growing from millions of years, into multiple millions into a billion year span and now into multiple billions of years. The biblical account of the actual years is often discounted by those that have adopted the time frame of evolutionists. Many Bible believing readers have become fearful that they will be ridiculed, if they do not accept the popular view of the old earth belief of billions of years.

The late Henry Morris is one of the most prolific authors and founder of the Institute of Creation Research is still recognized for his teachings for a young earth or more recent age of the earth. Morris declares that the best known chronological system based on biblical data is that of Archbishop James Ussher (1581–1656), who computed the date of creation as 4004 BC **(2).** Dr. Morris advocates that Jesus taught that every Word of God may be taken literally for three key reasons: (1) The Bible nowhere allows for long ages such as the Gap Theory. (2) The Bible explicitly states how and when creation took place. (3) The Lord Jesus recognized that men and women existed right from the beginning and even gave the names of those men in the earliest days of creation **(3).**

Some scientists that have suggested the billions of year theory for the age of the earth still rely upon the Carbon-14 (14C) dating system. This system consist of dating artifacts and biological materials; however, it does not hold a true test for things that are less than 50,000 years at this point. For those scientists that believe in a young earth, have long disregarded this test, since it has a host of inconsistent measures. In fact, the carbon-14 dating system has been flawed for many years. One of the more reliable systems being used today is luminescence dating technique, which measures the amount of radiation found in materials in the earth's crust. The longer the material has been in the ground, it is likely to have more radiation in the object. This measure may change according to the location of the object that is removed for analysis. Some soil areas may not possess full sun exposure, while others do, so the radiation will be higher depending upon factors such as time in full sunlight as opposed to only short exposure of sunlight.

Did Some Men Live over Nine Hundred Years?

According to the Bible, those first men lived long lives to propagate the earth with mankind. One needs to understand that there were no pollutants, no diseases, no high voltage power lines, no guns, no bombs, no airplanes, no trains, no automobiles, nor any heavy equipment. Until man was removed from the Garden of Eden, he was in a perfect environment with absolutely no dangers, nor threats to his life, whatsoever. There is evidence that God planned for man to populate the earth one person at a time from only one person, Adam.

While God filled the earth with multiple species of animals in mere minutes. The biblical account reveals that the seas were filled with the aquatic creatures and the heavens were filled in that same day with all species quickly. However, there were only 2 human beings prior to their expulsion from the Garden of Eden. God planned for all mankind to come through this one family, so when one accepts the biblical record that means that human beings have come through the family of Adam and Eve. The Bible account can and should be trusted as an accurate history of the events that transpired.

The record in Genesis 5 reports that the oldest man Methuselah lived to be 969 years and he died **(Gen. 5:27).** By the way, from the birth of Methuselah until the Great Flood there are exactly 969 years. This means that he either died in the year of the Flood or he died in the Flood. God had declared that if man rejected His righteousness and transgressed His Laws then he would surely die. Man failed God and committed sin that cost him his life. Even though, man did not die instantly physically when he sinned, man died spiritually at that moment of his sin by being separated from His Maker. The first man Adam lived 930 years **(Gen. 5:5)** and he died physically. Again, with no threats and in the plan of God to populate the earth from one man Adam, the longevity of life would become more effective for mankind to advance the population of man.

Modern science has an answer of one frequent area of debate about how did all the races of people come to exist if God only created one man Adam and all mankind came from that one man, how do we have all the varied races of people today? Science is still catching up with the Bible truths that man has been seeking. It continues to produce many of the answers that man has sought for thousands of years. The American Family Association reported in its monthly magazine in its September of 2017 that modern medicine has produced another answer to the human question of where did all the various races of mankind originate? Modern DNA studies have verified the connection from ancient mummies from ancient Egypt entombed in the Middle Egypt, discovered these mummies. Genetics match that of ancient people from the Near East, Turkey and the eastern Mediterranean area. The research was published on May 30, 2017, was done through Germany's University of Tubingen and the Max Planck Institute for the Science of human History **(4).**

These facts fit with the Genesis account from the Bible. "According to the Bible (concerning two of the sons of Ham) Mizraim settled in Egypt whereas Cush settled in Ethiopia, establishing two distinct and separate nations that did not share a common heritage **(5).**

There were 1,656 years from the birth of Adam until the Great Flood of Noah. The Bible records approximately about 2,400 years between the Great Flood and the birth of Jesus Christ. This means that approximately 4,000 years elapsed during the Old Testament stories until the birth of Jesus. If one accepts the biblical accounting of the years that people lived, then a student of the Bible can safely see that the earth is less than 10,000 years and likely just over 6,000 years for sure. This belief would often be referred to as a "young earth belief."

Was There a Global Flood?

Yes, the evidence is universal for a worldwide Flood of the proportions described in Genesis that formed the Grand Canyon, Victoria Falls (Africa) and reshaped the mountains around the globe. The foundational evidence would include the fossil records from every part of the earth, along with the records of fractures in the bottom of the seas. The ruptures of the earth's crust give plenty of evidence for the earth being broken open when the waters came spewing from the belly of the earth.

Paleoecologist Erik Gulbranson, professor at the University of Wisconsin at Milwaukee has discovered an amazing fossilized southern polar forest in a mountainous terrain near Antarctica. Gulbranson contends that the polar forest was formed over 260 million years ago. However, another scientist, Dr. Tim Clarey has written an article for Creation Research debunking the Gulbranson claim supporting the find is only a few thousand years old. Clarey based his research upon amino acids found in the fossils, which he states can only survive for thousands, not millions, of years **(6)**.

This discovery made by scientists would indicate the trees were buried rapidly during the global Flood described in Genesis **(7)**. These tree fossils were quickly buried in ash, mud and sand that engulfed them in an unusual cataclysmic manner. It further helps to explain the existence of animals, plants along with other live organisms were in abundance in each of the present-day polar regions prior to the Flood. These creatures and plant life existed in these polar areas prior to the Flood with a mild temperature zone all around the globe.

There are mountains of evidence in the waters cutting through rock formations with such force to form the canyons of the earth, plus the tonnage of water that brought the oceans level hundreds of feet higher than it had been prior to the flood. Fish fossils have been discovered high up in the mountains left from the global flood all around the earth. This clearly indicates that the waters covered the tops of the mountains, which coincides with the biblical account of **Genesis 7:19–20**. The Bible declared that the tops of the mountains were covered with water fifteen cubits (22.5 feet) or more.

The force of the water at this volume would have the power to cut through rock layers and form the Grand Canyon in a matter of days or certainly in a matter of weeks. Trickling water could not accomplish this amount of rock erosion over billions of years. These types of canyons around the earth were cut through rock layers in a brief period of time with a monster force of a huge flood in the day of Noah, there is overwhelming evidence of a worldwide flood scattered all around the earth.

Was There An Ice Age Associated with the Great Flood?

There is indeed evidence or an ice age that occurred immediately following the major cool down from the flood during the days of Noah. Scientists suggest during the time of Noah that as the water vapor canopy ruptured or the windows of heaven were opened producing sudden cooler temperatures throughout the entire earth. Much of the North and South Pole regions, perhaps even far beyond froze very quickly. Likely, areas such as Canada were completely frozen and even some of the northern parts of the United States were frozen following the sudden changes in the atmosphere. One might describe that time period following the Great Flood as "Global Cooling."

One of the problems that exist today is that so many scientists are unwilling to suggest the age of the earth is only slightly over six thousand years old, rather than so many that have believed the billions of years story. This accepted billions of years age has halted some scientists from attempting to develop their case for the Young Earth with the Ice Age that formed immediately following the Great Flood. Consequently, no one has come forth with a reasonable answer for how the Ice Age came to exist. Secular scientists have been unable to propose any new explanations, which is an indication that none of the current ones are convincing **(8).** Atmospheric scientist, Dr. Larry Vardiman used a standard meteorological computer models to simulate the effect of warm oceans on precipitation rates. His results demonstrated that very warm waters would dramatically increase snowfall in Yellowstone and Yosemite National Parks, which are known to have been covered by thick ice sheets during the Ice Age **(9).**

Some scientists have suggested that the Ice Age extended for three hundred to five hundred years following the flood, which kept the land bridges open for animals to cross over into other continents for that period of time. This period would explain how the animals reached around the earth as they repopulated the entire sphere of the earth. Once the huge sheets of ice begin to melt, the ocean levels began to rise and cover up the land bridges. One of the scientists that made this claim is R. D. Holt in an article published in the ***Journal of Creation*** in 1996. Holt states the massive levels of water following the flood began to freeze rapidly from a pleasant 72-degree temperature all over the earth to subzero temperatures toward each of the polar areas. The water stored in massive ice sheets would have temporarily lowered sea levels by 200 to 280 feet below today's level **(10).**

This same scientific data has also been published in *R. Wicander's Historic Geology* (**11).** There is enough compelling true science evidence of happenings after the flood, which is consistent with historical findings. The original creation was set in a perfect environment with the water vapor canopy in the upper atmosphere with temperatures somewhere around a pleasant 72 degrees all around the earth from pole to pole. Once that canopy was broken open and the sphere destroyed during the Great Flood, all the earth became radically different. The rich oxygen atmosphere became totally different, thus producing a much cooler temperatures immediately following the Flood. Scientists have suggested that this dramatic change lead to the Ice Age with a harsh, sudden downturn in temperatures. Some have suggested that it took a few centuries for the temperatures to normalize before most of the ice melted causing the ocean depth to increase a few hundred feet, as suggested by Holt and others.

During this new frigid temperature season, it became a different environment that man and animals were forced to face. While the animals were populating and moving around the globe, men were remaining in the Middle East and building the Tower of Babel. As the ice sheets began to melt, which again raised the ocean and covered the land bridges again, man was confronted with more problems. This huge development caused the end of much of the animal migration from one part of the planet to another as the land bridges disappeared, while the ice was melting.

The modern world of science has suggested other ice ages have appeared on earth over millions of years, which attribute these to the earth moving into varying orbit patterns. These scientists that have suggested such inconsistent movements of the earth away from the sun caused such ice ages. There is no evidence to prove these events have occurred, especially for the earth to move outside of its usual orbit pattern that the Lord placed it upon from creation. Many papers and various documents have been written on the subject; however, most of these have been proven to contain serious errors. Scientists believe that the great worldwide flood was so violent to the earth that it caused the earth to tilt about 23 degrees from its original position toward the sun. Thus, it is possible that this new alignment with the sun may have aided the temperature changes for a few years producing the ice age. It has been proven scientifically that the earth is currently aligned at a tilt of 23 degrees to the sun.

Did All Species Get on Board the Ark?

Again, the Bible has the Good News that God brought all of the animal species on the earth into the ark, so Noah never had to do the big roundup. Every animal that God wanted onboard the ark was on the ark. Many suggest there was not room for all of the species, but it was very possible. For one reply to their suggestion, the animals did not have to be full grown animals, as long as they were two of each kind. Many theologians have declared that it only takes a blue one and a pink one for the future of the species. In fact, nothing is impossible with the Creator of the universe.

The **Answers in Genesis** organization has built a life size Ark near Williamstown, Kentucky about 25 miles south of Cincinnati, Ohio which was completed in 2017. The newly built Ark has the biblical size given in Genesis which demonstrates how that this could have happened with over 1 million square feet of floor space inside the Ark. The exact replica of the ark now clearly illustrates for people to see the enormity of such a vessel. *Answers in Genesis* has allowed people to experience 3 complete decks of the Ark, which included all the thousands of animal species of the day that Noah went inside the Ark. According to publications as recent as 2014 estimates by the science world that there are less than 1.8 million species of all organisms. Around 98% of these species are fish, invertebrates, and non-animals (like plants and bacteria) **(12).** Some of the best estimates suggest fewer than 34,000 of these would be land dependent vertebrates in the present world today **(13).**

Perhaps, many of the aquatic species in the waters survived all of the threats that came against them. The Lord could have protected many creatures while, at the same time, He was destroying others. Rest assured that all creatures were right where He had planned for them to be.

Did God Create the World in Six Twenty-Four-Hour Days?

There is no evidence that would suggest that God did not create the earth, the heavens and all within them in nothing but 6 days with 24 hours each. If the days or the hours came from a mere allegory, then the real would have to proceed the allegorical period. Nothing is before the Genesis creation story to set the stage for an allegory. Genesis teaches that God put all the seasons together and there is no record of this period ever being changed. In fact, in Genesis chapter 8, it becomes abundantly clear that God's perfect work was established as He declared *that while the earth remains, seedtime and harvest, cold and heat, winter and summer, and day and night shall not cease* **(Gen. 8:22).**

No person can change the boundaries that God has set into place. There was at least one time that God, Himself changed the order of time, so that His army could win an important battle over an evil enemy. The Lord paused the position of the sun for a portion of a day so that Joshua could win the battle. That occurred in the battle with the Amorites in the Valley of Aijalon **(Joshua 10:11–15).** This was a supernatural work of God and does not happen upon its own or by the power of any man. God has fixed the times and seasons by His own Hand.

Who Made God?

The answer is simple, as the Bible contends that God always "was, is and will always be" nothing more nor less. Chapter 6 will discuss the Eternality of His Being, as He has always existed. There has never been a time that God did not exist. *"He is before all things, and in Him all things consist"* **(Col. 2:17).** Nobody made God, as He is the only Eternal Being of all times.

All other so-called gods of this world have been conceived in the hearts and minds of mankind, but not our Lord God. This invisible God that spoke all things into existence, must be accepted by faith. Faith is something that cannot be seen with the human eye, but something that one knows is true. No person was there to prove that everything happened as it is printed in the history of the world and the human race. Thus, the printed Word of God must be accepted by faith. The God that made this world, who is "the Beginning and the End, who is and who was and who is to come" **(Rev. 1:8)** as the Word of God declares. All mankind has faith in something or someone, so this is simply a matter of where a person chooses to place his faith in the reliability of the Bible.

Was There Another World before the Present One?

There are many people who follow the teachings of the **"Gap Theory,"** which teaches that millions or even billions of years could have existed between Genesis 1:1 and Genesis 1:2. Many of these say that there was a prehistoric period where the dinosaurs roamed the earth millions or billions of years before man. The Bible never suggests nor teaches anything like this. However, the Bible teaches that man and these huge creatures inhabited the earth simultaneously. There is no evidence of such a happening as the Gap Theory, so no the Bible declares this is the first earth.

A simplistic answer to this question is that nothing physical existed until God came from nowhere and spoke the worlds into existence. He merely gave voice commands and things came out of thin air to become solid objects. His invisible voice made things to suddenly appear from nothing. Man can make things from wood, metal, iron ore, soil, straw and other physical objects; however, man cannot manufacture something from nothing, but God did.

However, the Word of God plainly and clearly declared that God made this world from nothing and did not make it from physical material that already existed. In fact, the Hebrew word in the

Bible for create is *"bara,"* which can only be ascribed as the work of God. Another Old Testament Hebrew word that is only associated with the work of God is *"ex nihilo"* that is totally the work of God. The story is Genesis has to do with *"the Beginning"* and not a remake of something already in existence. This is a story of origins and science knows nothing about origins, as science is concerned with how things exist and has nothing to say as to how they begin **(14).** The creation account in Genesis is an actual account of God creating His world from nothing.

What Was the Earth Like before the Flood?

The earth was created in a perfect form with pleasant features for mankind. God carefully and meticulously constructed all that He made for the crown of His creation, which was man. God designed a perfect creation of earth, moon, sun, stars, heavens, all animals, plus a perfect man. All that God designed and manufactured was called Paradise, which means that it was impossible to improve upon God's creative masterpiece. When God completed the universes, there was heaven upon earth. Later, when God sent His judgment upon sinful man, He used the Great Flood to destroy the surface of the earth and spared one man and his family.

The Flood changed the entire surface of the earth, as well as, leaving some serious ruptures all around the earth. These fractures are found all over the surface of the earth today as reminders of a huge shift in the crust of the earth and in the deeps of the seas. These fractures in the earth are commonly called Fault Lines. The Flood brought death to most of the species of animals, human beings, while it reconfigured the entire crust of the planet. Chapter 10 will give much attention to this question and its answer.

Do Angels Really Exist and What Do They Do?

Yes, angels are created beings made by the hand of God to be His servants, as well as, servants for mankind, the highest creatures of God's order. There is an extended study in these materials in chapter

5 about God's plan and purpose for angels or messengers. Angels are real beings, but without flesh and blood, since they are spirit beings. On certain occasions in the Bible, they momentarily take on somewhat of a human form in order to deliver a message from the Lord to man. While they may take on that human form, they are not the same as humans. No man, woman nor child will ever become an angel.

Recently, an obituary column in a local paper carried this story of a sixty-five-year-old lady that died and whoever wrote the story declared the lady immediately became an angel at her death. The article further stated that the lady received a set of wings upon the same occasion and flew away to heaven. The writer did include the account of the lady having been a faithful member of a local church in her community. There are people that sincerely believe this story and many other stories that are not biblical. The Bible does include a truth that many may "have unwittingly entertained angels" **(Heb. 13:2)** without being aware that they were angels, since they had presented themselves as human beings. Angels may appear in these situations as human, but they never become men and men never become angels.

Jesus never became an angel in order to die for an angel's salvation. When angels sin against God, they fall without any redemption. Others such as the Sadducees raised a question for Jesus regarding the resurrection and marriage. Jesus clarified their belief on marriage in heaven that all redeemed would be brothers and sisters, while being "like the angels" **(Matt. 22:30).** Jesus simply declared that there would be no marriages performed in heaven. Apparently, these angels or "messengers" have never been granted the privilege of marriage that was only given to mankind. All the references in the Bible are always of the masculine gender with only angels named are male. Angels do not reproduce baby angels. They simple appear to be gender neutral as created beings with the purpose of serving God and man. Thus, Satan or Lucifer, who was an angel that committed sin and fell from heaven, will never to be redeemed according to the Scriptures **(Isa. 14:12–15**). He shall be in the lowest depths or at the bottom of the Pit of hell (**Isa. 14:15**).

Chapter 3

Frequently Asked Questions

1 ***Acts & Facts Magazine*, Institute of Creation Research,** Dallas, Texas, vol. 47, by Brian Thomas, from Thomas, B. Best, posted on Creation Science Updated from Dec. 2017, June 2018, pg. 20.
2 Morris, Henry II. ***The Genesis Record***. Baker Book House Company, Grand Rapids, Michigan: 1976, pg. 44.
3 Morris, Henry III. ***Acts & Facts Magazine***, ICR, July 2016, pg. 6.
4 American Family Association, ***AFA Journal***, Tupelo, Miss., September 2017, pg. 8.
5 **Ibid**.
6 **Ibid.,** March 2018, pg. 10.
7 **Ibid.**
8 Hebert, Jake, ***Acts & Facts Magazine***, The Bible Best Explains the Ice Age, November 2018, pg. 10.
9 **Ibid.**, pg. 11.
10 Holt, R. D., ***Acts & Facts Magazine***, (ICR), Evidence for the Late Cainozoic Flood/Post Flood Boundary, Dallas, Texas: August 2016, pg. 9.
11 Wicander, R & Monroe, J. S., ***Historical Geology***, 7th Edition, Brooks/Cole, Cengage & Learning, Belmont, California: 2013.
12 Answers in Genesis, ***Ark Materials***, Ken Ham, Williamstown, Ky.
13 **IBID.**
14 Phillips, John, ***Exploring Genesis***, Loizeaux Brothers, Neptune, New Jersey: (originally printed by Moody Press), 1992, pg. 38.

CHAPTER 4

A BRIEF ACCOUNT OF CREATION THEORIES

In chapter 13, there is a far more detail description of these modern-day theories along with some more ancient beliefs that people have held over the centuries regarding how all things came into existence. The first thing that any student of creation needs to know is that the book of Genesis is a credible account of the history of man and his beginnings. Genesis has withstood many examinations over the thousands of years by many renowned experts in biology, geology, archaeology, astronomy, paleontology, microbiology, ecology, philosophy, and many other areas of study.

One must either accept or reject the Bible as an authentic book of actual facts for themselves. Each person will choose for himself how he will view the Scriptures and the story of how all things came into existence. It requires faith, since no human was there when God created the heavens, the earth, the creatures and all that are a part of His initial creation. Man can take the authority of God and His only book that He wrote, the Holy Bible. Some will receive His Word, while others choose to reject the Word of God. A recent October 2018 Pew Research reveals that 60% of Americans accept at least one New Age belief-reincarnation, astrology, psychics, or a spiritual presence in physical objects. Forty percent believe in psychics and that spiritual energy is found in physical objects. Thirty-three percent believe in reincarnation and 29% believe in astrology **(1).** That means that at least 60% of all Believers in Jesus Christ as their Savior believe in at least one of the before mentioned beliefs. Many of these would be a part of Bible believing churches in America, meaning that they are in many, or most of Christian congregations. Pogo comic strip character stated that "we have met our enemy and it is us." It is true that many regular church going folks now have accepted one or more of the world views on creation.

The following theories are some of those that anyone beginning to study on the subject of creation most likely will encounter. The enemy always holds clever answers of how things came into existence without giving God any credit. Remember that these are the author's notes and not an in-depth study.

A. Agnostics: These proclaim that there simply is no god, thus god does not exist in their thinking. Some of these may be truthful; however, when pressure is applied to many of these in their personal lives, their story often changes. They would rather have others to believe that they believe that there is no god, than to have anyone to talk with them about the God of this universe named Jesus Christ. The Bible says, *"The fool has said in his heart that there is no God"* **(Ps. 14:1 and Ps. 53:1)**. Also, **Proverb 12:15** declares, *"The way of a fool is right in his own eyes."*

A Christian movie was produced in 2014 entitled, ***God is Not Dead*** by Sony Pictures, which portrayed an arrogant college professor that bullied and intimidated his students with his theory that God did not exist. He was so challenged by a student in the classroom that believed that God did exist that proved that his theory was full of holes. In the end the young college student proved to his classmates and to the professor that God did exist, even in the life of that professor. The professor had actually believed previously that there was a supreme God; however, because of some events in his life that had not turned out to meet his expectations, he denied God's existence. Throughout history, people have reported similar cases of people that tried to convince themselves and others that God is only a myth.

Some soldiers have declared that there are no atheists in a foxhole in middle of a heated battle. There are people that deny the existence of God to keep Christians from talking to them about the God that they need to know, who created the heavens and earth. These use this idea of no god to keep from hearing or admitting that God really does exist.

B. **Evolutionary Theory:** This is the most commonly accepted theory in most of the world today in textbooks and in most classrooms. Some people that hold this view often refer to themselves as **"atheists."** The view is often widely accepted without any question even among children of Christian Believers. In countries like USA, there is often no other version being taught, since teachers either do not want to be reprimanded by their school boards and possibly lose their jobs or become ostracized for taking an opposing view. In some cases, teachers are often unsure of their personal rights in a public classroom as to whether they can present an opposing view to the textbook version.

Those that follow this view of creation conclude that things operate by random chance, which have brought all things into existence today. Certain things came together or happened at the same time in order that the next step happened. This idea of chance is like rolling a pair of dice to see what comes forth. Those that follow this line of thinking believe that all things evolved from something very simple and as time progressed have now become far more complex creatures. This theory suggests that all things are constantly improving and moving in an upward sort of spiral implying that the best is yet to come. The major objection to this view is that the whole concept is wiped away by the Laws of God or often referred to as the Laws of Nature. For example, the Second Law of Thermodynamics is one of these laws that states that things left to themselves will begin to fall apart rather than improve. Even a newly built home will soon need improvements, since things will not repair themselves.

Charles Darwin is often heralded as the father of Modern Evolution; however, the idea or theory actually began far before the arrival of Charles Darwin. Darwin was trained in theology before he departed from pursuing the truth of the Bible and began to write about his thought life. Some of the writings of Darwin, such as his most famous **"The Origin of the Species"** propose the thoughts that God used this process to form human beings from lower species of life. Darwin wrote extensively with the concept of a common ancestor for all creatures of the animal world as well as the plant world. The Bible contradicts the Charles Darwin theory in Paul's writing to the church at Corinth **(1 Cor. 15:39)** when he clearly stated that *"All flesh is not the same flesh, but there is one kind of flesh of men, another flesh of animals, another of fish, and another of birds."* While birds with feathers may flock together, they are not the same as human flesh. Jesus never came to die for a fish, a bird, an elephant, a cat or a dog, but He became a man to die for all mankind. His fleshly body was made of dust like all mankind. Paul further states his case that *"we shall also bear the image of the heavenly Man"* **(1 Cor. 15:49).** Every human must decide whether to believe the Darwin's of the world or the Bible for their source of absolute truth.

C. **Theistic Evolution**: This has become a view that some Believers have accepted in light of all the literature written claiming that evolution is a fact instead of a theory. In recent history, evolution was taught as a "**theory**," however, more recently it is presented as a "**fact**" and in fact, the word *theory* has been removed in most recent writings. A theory is merely an unproven thought in the mind of a man and basically that is what evolution is plain and simple.

Modern man with classroom exposure to a more scientific approach to creation wants to believe that God used the unproven scientific ideas of evolution to bring about all life on earth. These deceived followers have married the two ideas of pure evolution and the Bible account of creation, thus accepting the idea that God used the process of evolution to bring all living things into existence. There is simply no way that true Believers can accept this theory and the Bible account of the beginning of life. If this were true then, how did the first non-living thing receive life, which it passed along to others?

D. **Gap Theory**: This theory attempts to accommodate the Bible account of creation with the evolutionary theory by adding the billions of years that most of the science world claims that preceded the existence of modern man. Those that follow this thought line simply believe that millions or billions of years existed between Genesis 1:1 and Genesis 1:2. This could be called the "**accommodating theory**" of belief. It compromises the Bible account with an unscientific and an unproven thought. Some refer to this belief as **"The Big Lie,"** since it propagates the idea that evolution has already been proven with the billions of year concept.

E. **Big Bang Theory:** This teaching is related to the two previous theories with the millions and billions of years. It has the beginning of civilization coming as a result of a huge cosmic explosion resulting from some ancient meteors or asteroids crashing into one another forming the planet earth. Opinions differ on whether God was involved in this explosion. The Big Bang started the wheels rolling with this super energy, which continues to support the planets into their present-day rotations. Many followers of this view sometimes refer to themselves as agnostics or Non-Believers in a Supreme God as Creator. Commonly these that hold to this belief say that things simply lined up by blind chance with certain meteors colliding with planets, which created tremendous energy that brought the present creation into existence.

This theory is closely connected with agnosticism, which supports the concept of "no god." Many followers of the Big Bang idea suggest that things in nature just happened to line up in such a sequence that they created an all-powerful explosion that produced the planets and all living organisms. Their explanations or very sketchy and unscientific, which suggest mother-nature is in control. God the Creator is left out of their equation, thus people need not give credit to anyone, especially to a supreme being.

Oddly enough, this theory was hatched by a Roman Catholic priest named Georges Lamaitre in Belgium in 1923, who never served a local parish, conducted a Mass, nor heard any confessions. He claimed his interest was more in serving the cosmos and humanity.

F. **Intelligent Design Theory**: This is a more modern term that evolved just before the turn of the twenty-first century. It seeks to define creation by a super being without always referring to this individual as God. It generally falls short of naming the Being. Some actually say that is God, while others may say "a god." The followers of this belief certainly leave the door open for Christ and the teachings of the Bible, even though they often stop short of calling His name. Many of these believers oppose the idea of evolution and random chance theories.

These clearly contend that the worlds did not create themselves and that someone with a design made all things that have been made. Intelligence was definitely involved in bringing about the complex creation that exists today. It would be foolish for these to believe in the idea of random chance of man or any other work of creation. The idea is from the Bible and they may even say that their view is rooted in the stories of the Bible, yet they may not believe all the Bible is accurate. On the other hand some will believe the Bible account, while they wholeheartedly refute the idea of blind random chance. These generally understand true science from the non-scientific version found among evolutionists.

G. **Biblical Creation Story**: This view is still a very popular view that Christians and others have held over the centuries from the Bible story in Genesis. Unfortunately, many people of this culture do not focus upon the Bible and rather view it as an old book that is unrelated to the events in this modern world. In fact, the view is no longer taught in most public places other than at church. Even many churches have compromised the Bible account with the textbook thoughts of the era. This has sent a very unclear message to the younger generations today. Many churchgoers are unsure what they believe and thus have mixed several worldly views with the Bible account. There is a great need to teach with clarity and share the Bible account with those in, as well as, out of church for them to understand exactly what the doctrine of Bible creation declares on this subject.

Chapter 4

A Brief Overview of Creation Theories

1 ***AFA Family Journal.*** American Family Association, December 2018, pg. 5. (pewrearch.org, 10/1/18).

CHAPTER 5
THE PRE-CREATION STORY

Exactly, what did transpire in eternity prior to God's marvelous plan of creation unfolding? Was there a bud before the flower of creation appeared? There are thousands of thoughts written upon the subject, but since no man was there to report it, many are forced to rely upon speculation or theories. However, for the Believer, there is the Word of Truth, which gives bits, pieces, glimpses of the events and sometimes whole accounts leading up to the actual process of creation. Some of the most significant references are found in the Old Testament books such as **Isaiah 14, Ezekiel 28, Job 1–2, Psalms, and Genesis 1–3**. New Testament references that give evidence of Satan and the angels in pre-creation are **Revelation 9, 12, 20, Luke 10, 2 Corinthians 11, Colossians 1 and Matthew 25**. Thus, each inquiring mind must decide whether he can trust the biblical records or simply reply upon some theory of speculation for the events that led to creation.

The Bible clearly affirms over and over that God is a Spirit Being, but there are other spirit beings mentioned in the biblical accounts. A student of the Word of God should understand when these other spirits came into existence. Some often asked questions are: (1) Did God live alone before creation? (2) Who is Satan? (3) When did Satan and the angels come into existence? Most serious Bible students search to find answers to such questions. This writing will attempt to expose some of those teachings for the Bible students to those accounts that could be helpful in answering such questions. A Bible student needs to be prepared to search the Scriptures in order to discover the truths. When a person searches for precious gems and jewels, he should not anticipate finding these lying around upon the surface of the earth. The more precious gems are found deep down below the surface of the earth's crust and so is it with the truths in the Word of God.

1. Did God Live Alone before Creation?

In religious literature, writers often refer to this period of time as the pre-Adam or pre-Adamatic era. For a lack of better terminology, the period before actual creation occurred will be called the Pre-Creation story. There were many events that apparently occurred in eternity prior to the physical development and existence of this world. First of all, it needs to be clearly stated the God was always in eternity and there was a period in history where nothing but God existed. There were no people nor angels. Paul said, "He is before all things" **(Col. 1:17).** Sometime, somewhere God created angelic beings to fulfill a need for man and the earth.

However, there is no sound evidence that any of these existed prior to mankind as the Bible remains silent on such a time. The writer of **Hebrews (1:14)** states that *"angels are ministering spirits for those who will inherit salvation,"* which clearly declares that angels were created to assist mankind. The Bible does not go into detail as to when these creatures were created. Many theologians believe they came after the creation of man, since everything was placed into a perfect system, thus, these could have come following those days of creation or during that same period of time. God announced that all things were *"very good"* **(Gen. 1:31)**, which was before any mention of angels or fallen angels.

Remember that God does not need anyone, nor anything apart from Himself, so he had absolutely no need for any man nor angels. Thus, God lived throughout eternity in a self-sufficient manner without any needs. This is one of those profound statements regarding the God of creation. A serious student of the Bible should understand this great truth that God needed no one nor nothing. Thus, He certainly does not need any one of us; however, we all need Him. Although man may not be knowledgeable of all the sequence of creation, we may conclude that God created the angels around the time frame that He created the heavens and earth. This would include those 6 actual days of creation of mankind described in Genesis 1 and 2.

When one reads the creation story, it does not take long to see that God created everything down to the details for man prior to his arrival upon the scene. Man is used in this writing in the generic sense that refers to all mankind and does not attempt to devalue the status of women. God is always referred to in the Bible with the masculine pronoun and never as she or it. It is unfortunate that in the political correct world today, that some already have altered that truth to fit their own beliefs. These agents of change have done so at their only demise, according to the Word of God. A final warning in the book of **Revelation (22:18–19),** that anyone who adds to or subtracts from God's written Word will suffer severe consequences with the plaques found in the Bible, plus God will remove him from a place in heaven.

Since the Lord God paid so much attention to detail in His story of creation, then it stands to reason that the stage was completely set, when He introduced man to His world of time, space, and all the creatures. Prior to creation, time did not exist. God created time on earth to aid mankind as a tool for discipline. God continues to live in a state of eternal now, where everything that He desires is in a presence tense. Thus, He created time as a part of the universes and placed the laws of God in effect to govern over creation. Some people refer to these laws of God as the Laws of Nature. For example, the Law of Gravity helps to govern over the work of creation. This law keeps man fixed to the planet of earth, rather than be free floating around in space. God knew that man and all life would need these laws to govern man's movement upon earth. With perfect insight about human nature, the Lord God placed everything on earth that His creatures would ever need into a perfect paradise.

God's construction project of designing and speaking the worlds into existence is an exceedingly huge project for man's comprehension. However, no project for God is a big undertaking, since He is such a Big God. He simply wanted a creature that resembled Him in thought and thus God set the stage to bring mankind into existence. God allowed man even to name things upon the earth, as well as God's creatures (**Gen. 2:19–20**). Somewhere in pre-creation, God alone planned everything that He would produce and carefully arranged the sequence of events of creation without any help nor advice from anyone. God did not need any assistance then, nor now to develop anything new. He and He alone existed before anything physical or spiritual existed. However, the creatures that He manufactured all have a need for God. That deep genuine need will never be satisfied apart from their Creator.

2. Who Is Satan?

In chapter 3, the Genesis writer introduces Satan to his readers by declaring him as "the most crafty of all the animals that God had made" **(Gen. 3:1)**. Satan not only shows up on earth very soon in creation, but he is very different from man. The writer says he is of the "wild animals" by stating that he *"was more cunning than any beast of the field"* **(Gen. 3:1).** This creature was not a beast by God's design. If he got to earth in the Garden of Eden, then how did he arrive?

The writer of Job stated, "The Lord said to Satan, where have you come from? Satan answered the Lord, from roaming through the earth and going back and forth in it" **(Job 1:7).** This piece of information tells the Bible student that Satan does not possess the power of God to be at all places at the same time. He has to travel from place to place as a spirit being, which differs from an animal, but resembles a beast in the book of Revelation. God's question for this subtle beast of the field indicates that God is asking him to confess up to his original design for this deceitful beast. Genesis 3 could have been many years after those 6 days of creation or just a few days following. The Bible is silent on the time period.

Other names of this beast of the field are "the dragon, Satan, that "ancient serpent," and "the devil" **(Rev. 20:2).** All of these seem to possess the same devious character of deception and carry the same idea of perversion of the truth. The Bible, also calls him the father of liars **(Jn. 8:44).** In the very same book of the Bible, Jesus describes Himself as the Way, the Truth, and the Life **(Jn. 14:6).** In these accounts, it is clear that Satan and the Lord Jesus are complete opposites in purpose and mission.

Devil (diabolos) means a false accuser, Satan, and slander in the Greek and is the word from which the English word diabolical is formed (**1**). Here are some of the New Testaments references given to this devious character: He is called "the tempter" **(Matt. 4:3),** "Beelzebub" **(Matt. 12:27),** "he is a liar and the father of lies" **(Jn. 8:44)**, "the prince of this world" **(Jn. 12:31)**, "the god of this age" **(2 Cor. 4:4),** "Abaddon," which means destruction, **(Rev. 9:11)**, "serpent of old" **(Rev. 12:9),** and "the dragon" **(Rev. 20:2)**. The final mention of the devil is found in **Revelation 20:10**, after the thousand-year reign of Jesus the devil is cast into the lake of fire and brimstone with the beast and the false prophet to be continuously tormented forever and ever. He is no more in charge of anything, nor will anyone have to deal with him again.

According to Ezekiel **(28:12–19),** the devil was created *"perfect in your ways from the day you were created, till iniquity was found in you"* **(Ezek. 28:15).** Ezekiel declares that the devil was an anointed cherub **(v. 14)** that was cast out of the mountain of God to the ground because of his sin and one day will be burned to ashes. According to **verse 17**, *"this heart was lifted up because of his beauty."* **Ezekiel 28:16** states this truth, *"By the abundance of your trading you became filled with violence within, and you sinned."* Of all the things and beings that God created, this was the first creature to ever pervert God's perfect world. This account is very consistent with the entirety of Scripture from Genesis to The Revelation. Satan showed up in the Garden of Eden **(Ezek. 28:13)** while Adam and Eve were living in the garden as the devious character that tempted Eve to sin, which meant that he had already been removed from heaven.

The Prophet **Isaiah** said, *"How you have fallen from heaven, O Morning Star, son of the dawn! You have been cast down to the earth, you who once laid low the nations! You said in your heart, I will ascend to heaven; I will raise my name above the stars of God: I will sit enthroned on the mount of assembly, on the utmost heights of the sacred mountain. I will ascend above the tops of the clouds; I will make myself like the Most High"* **(Isa. 14:12–14).** These statements from the mouth of this devious creature are often referred to as his famous five "I wills," which will be discussed shortly. It was all about his personal will over all other creatures and the Lord God.

There are some great truths in this passage that one must be careful not to bypass in reading. First, the devil was in heaven before his fall, so he already knew of the beauty and majesty of God's wonderful heaven and had even seen the Lord's throne chair. From this observation, one can certainly understand that Satan is a spirit being and was excommunicated from the bliss of heaven, which he had previously enjoyed. However, the intent of the devil is clearly stated to one day climb back up to the lofty tops of the mountains of heaven and seat himself upon the throne of God. This reveals serious rebellion of this beast which is identified in the New Testament as the antichrist. The idea of the antichrist opposes all that Christ stands for and that he is a phony or a counterfeit. He chose to oppose God, and willfully lead a rebellion to overthrow the Lord.

This shadowy creature always shows up early in every person's life to thwart the will of God in each individual's life. He showed up early in the Garden of Eden to the first Adam and then early for the Second Adam-Jesus in the wilderness, while He was meeting with the Father in preparation for His earthly ministry. This indicates the boldness of his character. Satan seemed to carefully watch and wait for the most opportune time, when the flesh of Jesus was weak and depleted. Jesus had been fasting and praying for forty days when Satan approached the Lord, in order, to sabotage His ministry. The gospel writers of **Matthew (4:1–11), Mark (1:12–13), and Luke (4:1–13***)* all record the temptations of the devil upon Jesus. Even though, Jesus was weak physically, He demonstrated to all Believers how to resist the devil, by quoting Scriptures which the devil could not deny.

God has a definite plan for the life of every creature **(Jer. 29:11)** and so does the false Christ. He has set out upon a mission to destroy, whereas, Jesus is upon a mission *"to seek out and to save"* **(Jn. 19:10)** those sinners that are without a Savior. Again, these are contrasting courses of action that places the devil at the opposite pole from Christ. Satan is then referred to as the antichrist that opposes God and all good. Therefore, make no mistake that this evil creature is an enemy to every human being and not a trusted friend to anyone.

Satan's name, Lucifer, means "shining one," "bright one," "glamorous one," and certainly Satan could be the most beautiful creature that ever came from the hand of God! **(2)** Paul declared that "he has the ability to transform himself into an angel of light" **(2 Cor. 11:14).** Isaiah declares that Lucifer is guilty of the five terrible "I will's" against the will of God. (1) The first "I will" ascend into heaven. Satan is speaking of the third heaven where the throne room of God is located. The first heaven is that visible sphere around the earth with all the sun, moon and stars. The second heaven is the abode of the angels beyond the first sphere, where man cannot see. The third heaven is the abode of God. Satan based his will to go to the third heaven on pride, because he has an ego problem. (2) The second "I will" is based upon an unholy ambition of power to rule over the angels. (3) The third "I will" expresses a desire to rule over the inhabitants of the earth as he says the assembly. (4) The fourth "I will" sit above the clouds and take over the entire universe. (5) The fifth "I will" desire was to be God and receive all worship.

Consequently, due to his ambitious will to pervert, Satan must experience three different falls. The first fall occurred in Isaiah 14, when he fell from heaven. The second fall is described by Jesus in **Luke 10:18** as He said, *"I beheld Satan as lightning fall from heaven,"* which represents the fall in the Tribulation period. The third fall is the casting into the lake of fire at his final judgment **(Rev. 20:10).**

Every Believer can take heart in Isaiah's message, because the devil who roams the earth will be brought low to his grave called a pit **(Isa. 14:15).** This arch enemy of God's fate has already been sealed, whether he believes it or not. The reference to the pit is the same bottomless pit in **Revelation 20:1.** Since the cross, Satan is squirming like a snake that has received the mortal blow, but is still alive with activity. Sometimes man can even cut the head off the snake, but the body continues to twist and squirm for a long period of time. Jesus crushed Satan's head at the cross, but the body is still alive squirming to wreck other lives and to deceive everyone that he can. This is consistent with the punishment for the devil's deception of Eve in the Garden **(Gen. 3:15),** which is the first pro-evangelism verse with promise in the Bible.

Another heartwarming bit of information for every Christian is that the old devil is done away with in **Revelation 20:10.** He along with all the wicked will be forever sentenced to the eternal lake of fire of torment day and night. Satan is not in the first two chapters of the Bible, nor is he in the last two chapters of the Bible. Praise the Lord! The righteous children of God will not be subjected to his deceit, lies and vicious attacks after Christ's final judgment upon him. Isaiah's message was one of hope for the righteous, but a message of doom to the kings of Israel like Ahab who were leading open rebellions against the Lord. They too would suffer the consequences of a trampled and fallen leader like their father the devil **(Isa. 14:19–20).** The Lord would give the world a prelude of His final judgment as He dealt with the rebellious people of Israel.

The Bible contends that Satan is an adversary to the work of God. The Lord God did not create evil, but He did give an opportunity for the man created in *"the image of God"* **(Gen. 1:26–27)** a choice. The Bible uses the word *"image of God,"* which means that man physically does not look like the Creator as much as, the likeness of thought and decision making privileges. Thus, man is endowed by his Creator with special abilities unalike any other creature on earth to possess a thought life. With this privilege comes tremendous responsibility to the Creator, which is taught in the Bible.

In summary, one may say that the devil was convicted of his sin of rebellion while living in Heaven and was then sentenced to a temporary domain upon earth. After the destruction of the earth, the Lord has assigned the devil to the pit of destruction. Even in the pit,

Satan will not be the ruler, but a prisoner in a pit of torment for all of eternity. God makes no exception for a sin filled life. Apparently, there is no suggestion in the Bible that angels or cherubs will receive redemption. There is a song that even the angels cannot sing and that is the song of redemption. Thus, if an angel rebels against God, there is no prescription for redemption given in the Word of God.

3. Why Do Angels, Satan, and Other Beings Exist, and When Did They Come into Existence?

The third question deals with why are the angels and Satan needed and when did they appear in creation. Once again, this understanding can be found from the Bible, since it is always the primary source. One must return to the old, old story of creation with the understanding that God had put everything into being around the time He created man, so that man would lack for nothing. Man was God's crowning jewel of creation, which was created in God's own image.

It is interesting that the Bible never records the concept of "the image of God" with the angels nor with the animal kingdom. That teaching about "the image of God" is a reference attached only to mankind. Yet God *"crowned man with glory and honor"* **(Ps. 8:5),** which is never attributed to anyone except man. This great distinction is made in the teachings of the Word of God, since the Creator set man in a superior position above animals and angels.

Hebrews 1:14 describes these created beings as ministering spirits sent forth to minister for the benefit of human beings. Many popular books, articles, television programs and movies have been produced about angels. Unfortunately, many of these are fictional characters that have very little in common with those of the Scriptures. For example, nowhere in the Bible does it describe angels as creatures with wings. Only the cherubim and seraphim's are described with wings in the biblical teachings. Many people believe that they will gain wings when they go to Heaven, but this is not taught anywhere in the Bible. Others actually believe that they will become angels in heaven, but this is not true either. Again, the Bible does not state that people will become angels.

However, the writer **Matthew (22:30)** does say in one comment that man in heaven will become "like angels." This statement comes at the end of a discourse that Jesus had with some Sadducees, who contended there was no resurrection from the dead. They asked Jesus a question, "whose wife would the woman be who had seven husbands on earth?" Jesus answered by declaring that there is no marriage in heaven, but that people would be like angels without marriage. Unfortunately, many writers have deceived people about the appearance of angels. The best way to understand their appearance is to look at the people in the Bible accounts that actually saw angels. In practically each case, the person seeing the angel fell immediately into a state of serious fear. The angel would respond to that person's state of fright, by speaking the phrase, "Do not fear."

According to the Bible, angels are always of the masculine gender. For example in **Revelation 10:1**, John describing an angel that he saw, *"His head, his face was like the sun, and his feet like pillars of fire. In the next verse, he had a little book open in his hand. And he set his right foot on the sea and his left foot on the land."* Notice that John continued on and on by using the masculine pronoun to describe to his readers the gender of these creatures. Thus, angels being all of the masculine gender would not be married as humans.

According to the Apostle Peter (**2 Peter 2:4**), when an angel commits disobedience with God, he becomes a fallen angel and cannot be restored through any means. These fallen and disobedient soldiers became the demons that often afflict and try to persuade men on earth to commit sin. In the final analysis, Peter declares that they will be bound in chains of darkness and delivered to the judgment of God. It is apparent that these fallen servants of the Lord possess a leader that is known to the world as the king of darkness.

Notice that the Bible gives the name of the leader of those fallen angels. Their king was called Lucifer by Isaiah, but by John **(Rev. 9:11)** is referred to as the angel of the bottomless pit, whose name in Hebrew is Abaddon, but in Greek he has the name of Apollyon or the Destroyer in many translations. Isaiah declares that he was fallen from Heaven and was cast down to earth. He further says that this king of the fallen angels will be cast into the lowest depths of the Pit or hell. In the end, Satan will have no domain, nor authority over anything or anyone. His kingship will have an abrupt ending. **Isaiah** further states that people will gaze upon the fallen king and say, "Is this the one who made the earth to tremble?" **(14:16).**

Perhaps, it would be safe to conclude that God was in the business of preparing creation for man long before He spoke everything into existence. As a part of that plan in pre-creation, He could have created messengers and servants often referred to as angels, but for what purpose? It seems more likely that these angels were created sometime after the creation of man. The Bible describes angels as ministering spirits with the writer of **Hebrews** declaring an innumerable company of angels **(12:22)**. He further states people have actually entertained angels while believing they were mere strangers **(Heb. 13:2).** In **Hebrews 1:14**, the writer asks a question, "Are they not all ministering spirits sent forth to minister for those who will inherit salvation?" Here it becomes apparent that these angels are not flesh and blood, but rather spirit beings that were created to benefit man. Man is the only creature that will inherit salvation. Likely, they could have come into creation when man had a need after man's fall in the Garden.

According to the **Genesis**, God spoke after day 6 was complete by stating that **"it was very good" (1:31).** This would indicate that Satan, the evil one was not upon the scene, indicating that the angels were formed following the creation of matter. The heavenly host are all Spirit beings and not physical. This seems to be an indication that this creature was created after the Paradise on earth with man. Otherwise, God would have pronounced Satan's rebellion very good; yet throughout Scripture, God is absolute that sin is detestable in His eyes **(3).**

References to angels are all over the Scriptures. **In Hebrews 1:13**, the writer says that the Lord never invited an angel to sit at the right hand of the Lord and that angels are not equal to Jesus. Even though, the Bible does declare that Jesus came into His world a little lower than angels, thus speaking of His earthly incarnation **(Heb.**

2:7, 2:9). However, Jesus would suffer things that angels would never experience nor understand in God's plan of redemption. Jesus came into His world in the lowest of ways, but would be lifted up above all creatures and all beings that He had made. The Hebrew writer is making a case here that Jesus is Superior to all creatures, even heavenly ones. Notice that Satan was created with beauty in heaven, but after his rebellion, he has been relegated to earth, which is lower, but eventually to the lowest part of the pit of hell. Thus, the old devil started his journey as a high official for the Lord, but at the end will reach the lowest possible level for any creature that God ever made into the lowest part of the pit.

There is no indication of any exact number of angels. The Bible does not suggest that there is an angel hiding behind every bush or big rock, but there are plenty of them to do the work of the Lord when needed for God's people. They make hundreds of appearances at the command of the Lord and in specific situations to deliver messages to human beings, thus in the Bible they are heavenly messengers.

There was one rotten apple in the barrel named Lucifer or Morning Star, who led an open revolt against the Lord in Heaven. As a consequence of his rebellion, he was placed on earth as a part of God's permissive will and design. Satan is unable to force man to choose, so he is left up to his deceptive practices of persuasion, perversion, and lies.

Without wasting any valuable time on earth, the devil using the body of a serpent, approached God's first of mankind, and set out on a course to get even with God. By lying and making insinuations about God that were totally untrue, Satan gained an immediate victory in his new environment by ensnaring God's people. The Bible declares that "God cannot be tempted by evil, nor does He tempt any man" **(James 1:13)**. On the other hand nothing is honorable for the devil, so he tempts man to commit sin. He has no power to enforce his will upon man, plus, every temptation of the devil must pass through the hand of God. Therefore, God permits things to happen to all people, in order that man will depend upon God, instead of himself.

Perhaps, the obvious needs to be pointed out that there was no devil and no evil in the first two chapters of Genesis. This perfect creation of God was truly paradise, prior to the Fall of man. Nothing but good existed until chapter 3 when the evil one suddenly appears in the Garden of Eden, which was a perfect place with perfect peace. All does change immensely when the evil one arrives to suggest nothing but lies to Eve about what God had done. Evil had absolutely no place in the perfect Creation of a Perfect Creator. The Bible is clear that the devil had not been on earth until Genesis 3, when he suddenly appears to Eve in the Garden.

God set the stage for man to become the ruler of His creation **(Ps. 8:6)**. However, not long after the creation of man, Satan appears on earth to challenge God's people and ultimately, the Lord Himself. In God's infinite wisdom and prior to creation, he designed a plan for man who He created in His own image. That plan included man having the right to choose to love, follow and obey his Creator. However, if he was to have a choice then, he could reject God and live his life to suit his own selfish interests. God wanted man to choose Him with the straight and narrow path of life that would lead to eternal blessings.

If man were to choose, then there had to be a choice to be made. Since the Lord God possess perfect wisdom to know the outcome and results beforehand, He was willing to pay the price to obtain the love of His creatures without any manipulation on His behalf. If God were going to be fair and impartial in man's choice, then there had to be another option for man to select. So God gave a choice to His angels of heaven and their leaders. When that alternative was granted in heaven, an angel called "Lucifer, or Morning Star" **(Isa. 14:12),** rose up in sinful rebellion to receive the worship that belonged to God. In the story of the dragon in Revelation, there was a great battle in heaven. The dragon was able to gather up a third of the host of heaven, which were cast out of heaven to earth *(Rev. 12:4)*. The Morning Star was cast out as a fallen star to earth, where he was given the key to the smoking pit **(Rev. 9:1–2).**

According to John's account, the dragon was permitted to torment the children of God, but would be unable to kill them **(Rev.**

9:5). This story seems to parallel the happening of Job when Satan came for his second assault on the man of God. "The Lord said to Satan, very well then, he is in your hands; but you must spare his life" **(Job 2:6).** John also called the devil *"the accuser of the brethren, who accuses them before our God day and night, has been cast down"* (**Rev. 12:10**). Satan never gives up with his vicious attacks upon the children of God that continues day and night. This would indicate that while God's people are asleep and doing nothing amiss, this liar continues a relentless onslaught. A child of God might expect being attacked by the devil for simply being a child of God.

Satan has atta*cked* God's children from the very moment that he first arrived on earth. That means that Satan's rebellion came after man's creation and his opportunity to choose. John contends that Satan was not only a liar and the father of all liars, but that *"he was a murderer from the beginning"* **(John 8:44).** Since this is the same term John used *"in the beginning..."* **(John 1:1)** to refer to the earth's origin, then Satan was there shortly after the beginning of creation like a wolf licking his chops to challenge God's first man. Some mistakenly believe that he was cast from heaven in **Genesis 1:2**. Since Satan is the essence of darkness, then it was his presence on earth that brought an outer darkness upon the face of an empty earth where there was no light and thus complete darkness reined.

However, the Bible declares that all of creation was made perfect, which means that sin had not entered into the world at the time of the creation of Adam. After the creation of man, he began naturally to worship and adore his Creator causing the evil one to become jealous. It would appear that very little time elapsed before Lucifer rebelled and was cast out of Heaven. This means that the very first sin happened in heaven sometime prior to man's fall in the Garden of Eden.

Those angels who followed Satan's leadership became the demons of the earth that continue on a destructive path of conflict with the inhabitants of earth today. Man is taught not to fear the devil or his demons of destruction. "*For greater is He who is in us than he who is the world"* **(1 John 4:4).**

In **Job 38:1–7,** God Himself speaks to Job and asks him, *"Where were you when I laid the earth's foundation?"* He continues by stating that *"all the sons shouted for joy"* when the foundation was put in place. From this account, it is not any mention of angels or when they were created. It would clearly appear that the Lord created the heavenly host of angels, cherubim and seraphim about the same time of the six days of creation. Even though, Satan took a host of fallen angels with him, there are still enough angels to serve God and man today in whatever manner God deploys them for service.

The Believer is promised by God Himself that He will never permit any temptation that is too big to be brought before him, *"without making a way for man to escape that temptation"* **(1 Cor. 10:13).** It is important to note that this promise for the Believer is not available to non-believers, but only those that have received a personal relationship with Jesus. Satan's ploy in temptation is to cause a child of God to stumble and fall. However, the plan of God, when He permits a temptation is to cause His children to stand firm and tall in the face of the temptation. Since Satan is a longtime adversary of God and His children, man simply needs to accept the facts, while learning how to deal with the devil. He can be overcome in the lives of Christians through Jesus's name and blood. Truly, there is power in the blood of Jesus to overcome the evil traps that Satan sets for every Believer.

Another puzzling question asked by many Believers is, "Why did God create an evil being such as Satan?" The very nature of God declares that He is 100% good and that it is impossible for Him to do, to think, or to create evil. Apparently, God gave the angels a one choice plan without redemption for any that made the wrong choice. One may think that the Creator was unfair; however, He created only mankind to have eternal fellowship with and not the angels. Christ did not die for any angels, but died for the souls of the human race and not angels nor animals. It is all about the shedding of blood that brings a person, not an angel that has no blood into a right relationship with the Father. This would be consistent with **2 Peter 2:4,** when Peter declared that God did not spare the angels who sinned, but cast them down to hell (Tartarus) and delivered them into chains of darkness, to be reserved for judgment.

There is no blood atonement for the angels if they choose to rebel against the Lord God, but theirs is an eternal judgment of condemnation. Remember the Bible states that the angels saw God's face and all His heavenly glory **(1 Tim. 3:16).** These spirit beings were created by the Lord, placed in Heaven and around the throne of God to see firsthand all the glory and majesty of our Lord. If they could see what they saw, experience what they experienced and then rebel, they did not deserve to return to Heaven. Thus, it is impossible for fallen angels to be saved from their sin of rebellion. Christ had to become a man in order to save sinful man and become like man. For Jesus to redeem the angels, then He would have to become an angel to save angels. No such plan has been provided according to the Bible.

In conclusion, the first sin did not happen on earth, but it happened in Heaven by Lucifer and the host that followed him, thus God cast them out of Heaven. God did not create evil, but fully intended for each heavenly messenger and all of mankind to love and accept Him and His plan for their lives. Evil came about through acts of perversion by Lucifer and then man. The Bible teaches that there is nothing but good that comes from God. In **James 1:17**, it declares that *"Every good gift and every perfect gift is from above, and comes down from the Father of lights, with whom there is no variation or shadow of turning."* No, God is not responsible for anything evil, thus all sin can be attributed to either fallen angels or fallen man.

Chapter 5

The Pre-Creation Account

1 Hodge, Bodie, ***Satan and the Serpent***, Answers in Genesis Series, A Biblical View of Satan and Evil, Hebron, Kentucky: 2014, pg. 10.
2 Thieme, ***Creation, Chaos and the Restoration***, pg. 3
3 Hodge, ***Satan and the Serpent***, pg. 24.

CHAPTER 6

THE ETERNALITY OF GOD

(The Creator of the Cross)

In order to have a clear understanding of the creation of the world, one should be introduced to the Creator of the universe and learn as much as possible about His character. The Bible declares that God existed before the creation of the world. There was nothing prehistoric, but there was a preexistent God, since God has always existed. If one could travel back into past history and trace down the genealogy of man to his origin and then continue to travel on a backward path beyond man, matter, and time, there was always God. Often people have a difficult time putting the Bible teaching together and explaining this phenomenon of events regarding the Lord.

Here are two simple Bible thoughts that might help the complexity surrounding man's understanding of this eternal Being. First, Jesus Christ existed before his birth in the stable of Bethlehem a little over two thousand years ago. Second, it was Jesus as the Creator, who spoke all of the world and matter into existence just over six thousand years ago. Perhaps, there is no clearer picture of these two statements of fact than from the pen of the Apostle Paul in his writing to the church at Colossae with these words. Jesus *"is the image of the invisible… God, the first-born over all creation. For by Him all things were created that are in heaven and that are on earth, visible and invisible. He is before all things, and in Him all things consist.* ***(Col. 1:15–17).***

Many theologians suggest that Jesus as the pre-incarnate Christ, which made several Old Testament appearances with some invisible and others visibly much like a human being. Some theologians refer to these angelic or human appearances in the Old Testament as a theophany, which usually means, a visible manifestation of God to a human being. The Bible describes some of these when God began to speak to Moses through a burning bush down near Egypt as *"the Angel of the Lord"* (Ex. 3:2). Jacob had a strange encounter with the Lord prior to his return to the land of Israel, when he was about to meet his brother Esau. Jacob wrestled all night with a Man, whom led him to declare, *"For I have seen God face to face and my life is preserved"* (Gen. 32:30). Others who either talked with *"The Angel of the Lord was with Joshua in* (Joshua 5:13–15 and was with Daniel in the fire as the fourth Man walking with Daniel. (3:25).

Before anything else existed there was God, who is the very same God today that is in charge of the universes. The **Psalmist** declared in **90:2–4,** *"Even from everlasting to everlasting, You are God. For a thousand years in your sight are like yesterday when it is past, and like a watch in the night."* Some popular religions today, teach that the present God was once a man and lived an exceptional good life upon another planet. Beliefs such as these are in error, since they attempt to discredit the Lord as Creator of all things. God has always existed and was preexistent to the creation of the universes. Before one can really appreciate what took place in creation, he should get acquainted with the Master behind the creation events. There are several eternal aspects to God's character that every Bible student should possess a clear understanding of this incredible Maker of everything.

Throughout the Bible many of the writers understood that the Lord God was an Eternal Being. In **Psalms 102:25–27**, the writer declared the awesome glory of God, *"Of old You laid the foundation of the earth, and the heavens are the work of Your hands. They will perish, but You will endure or continue." Yes, they will all grow old like a garment; Like a cloak You will change them, and they will be changed. But you are the same, and Your years will have no end."*

God Is an Eternal Being

Children often ask honest yet revealing questions such as "Who is God?" "Who created God?" and "When did God come into existence?" The answers are often difficult to explain for an adult, but yet most often children can readily accept the biblical facts. First of all, God has always existed as He is today, without any changes in character, nature or being. God has not grown any older, nor become feeble. Rather than stating God is pre-historic, He pre-existed everything and everyone. No one ever created God, since the Bible reveals that He is, He was, and He is to come **(Rev. 1:4).**

An often debated question in this modern world is why are there so many different names given to God and which of these is correct. Some today are suggesting that this is confusing to people, who are demanding for all religions to come forward with a name for God that all religions can live with. One suggestion is that Christians join with the Muslims and others, who also believe in a monotheistic god to agree upon calling him Allah. They say that all are speaking about the same god. The Muslim religion, however, denies Jesus being the Son of God, since they believe that God had no Son. This Muslim claim represents a head-on collision between the God of the Bible and Allah **(1).** They also deny the presence of the Holy Spirit. A court in Malaysia recent ruling restricts any religion using the title Allah, except Muslims. No other religion has the idea of the Trinitarian view of three in one idea of the Christian theology. The name of God cannot be a trivial matter for the true Believer. The Bible has already given man the name of the True God. Jesus said, "*If you had known Me, you would have known my Father also*" **(John 14:7).**

"*Before the mountains were brought forth, or ever You had formed the earth and the world, even from everlasting to everlasting, You are God*" **(Ps. 90:2).** Thus, before the world was called into existence, God lived face to face with Himself in perfect harmony. He had perfect peace, perfect love and perfect sovereignty without any problems. The Bible says that *"before the world was that Christ died for us"* **(Rev. 13:8).** The Bible student says, "How could that happen?" In the heart of a great God who had carefully and thoughtfully designed the world and thus, committed Himself to saving sinful man, gave His own life upon the cruel cross.

In **Isaiah 45:18**, the Bible says, *"For this is what the Lord says-He who created the heavens, He is God; He who fashioned and made the* *earth, He founded it; He did not create it to be empty, but formed it to be inhabited—Isaiah says: I am the Lord, and there is no other."* It was apparent to Isaiah that God was one of a kind with no other equal. He had always been God and He would continue as the one supreme God. It is important to note that the Lord made this statement about Himself to the prophet. Some religions offer a different god for a different generation, like the kings of some nations of the past. They believe that they come and go like the seasons. The God of the Bible does not change, as *Jesus Christ is the same yesterday, and today and forever* **(Heb. 13:8).**

Nehemiah **(9:6)** says, *"You alone are the Lord; You have made heaven and the heaven of heavens with all their host, the earth and all things on it, the seas and all that is in them, and You preserve them all."* God superseded all things because His eternal existence is one of a kind. Nehemiah clearly and simply said that *"You alone are the Lord."*

The Psalmist declares that God's existence is eternal when he said, *"Your kingdom is an everlasting kingdom"* **(Ps. 145:13)**, while *"man is like a breath and man's days are like a passing shadow"* **(Ps. 144:4)**. Perhaps, the clearest revelation by the Psalmist is stated, *"from everlasting to everlasting, You are God"* **(Ps. 90:2).**

God Is an Eternal Triune God

The serious student of theology says, "Then did Jesus exist back there before the world?" Yes, Jesus had already suffered, bled and died for all of man's sins before God spoke the first fragment of creation into existence **(Rev. 13:8)**. That means that the Triune God of the Father, Son, and Holy Spirit were there before the world, united in One Being with One Purpose. Yes, the three in one are eternal in existence. God in three persons is a Spirit being and does not possess flesh and bones. Jesus came to live upon the earth in an incarnated body like all other men for his earthly purpose and pilgrimage. The Bible declares that *"Jesus became flesh and dwelt among us"* **(Jn. 1:14).** Otherwise, the God of the ages is a Spirit and those who worship Him must worship Him in Spirit and in truth **(Jn. 4:24).**

W. O. Carver pointed out, the Bible does not have a formal doctrine of the Trinity; it nowhere discusses the Trinity as such. It speaks of the Father, the Son, the Holy Spirit in a unity, but each of the three with distinctive characteristics and function **(2)**. The Bible does speak of the Father, the Son and the Holy Spirit to give an understanding of God, but it does not teach that these are three Gods. These teachings are principally found in the New Testament, yet as one comprehends that message, the Old Testament comes alive with the Triune God concept.

Any Bible study should be careful not to dissect the God of the ages into three parts, nor to mislead others in thinking that there are three Gods. Mark, the writer, says it rather succinctly when he recorded *"The Lord our God is one"* **(Mark 12:29)** and the Apostle Paul declared it well by stating, *"One Lord, one faith, one baptism"* **(Eph. 4:5).** The Apostle John said, *"there are three that bear record in heaven, the Father, the Word, and the Holy Spirit; and these three are one"* **(1 John 5:7).**

When John the Apostle introduces Jesus in his Gospel of John, he declares that Jesus is the Word as John uses the text of Genesis 1:1. As John gives an account of Jesus as the Divine Son of God that was incarnated into flesh, he refers to Jesus as the Word. In the Greek language that the New Testament was written, John calls Him the Word or "Logos." Many early religious leaders declared that Jesus was merely created, rather than being the Creator. Thus, this would have made Jesus less than Divine and merely a man.

Actually the "Logos" was a title as the revealer of God the Father. In **Colossians 2:15,** Jesus is referred to as the very image of the invisible God or one might say that *"Jesus is the visible image of the invisible God."* Jesus would say to his people, *"If you have seen me, then you have seen God."* Paul further declares that *"for in Him all things were created, in heaven and on earth, visible and invisible, whether thrones or dominions or principalities or authorities—all things were created through Him and for Him"* **(Col. 2:16)**. Jesus fleshed out that invisible into human flesh, so man could look upon the great God of glory.

The writer John in that first chapter explains that in the beginning was the Word, and that the Word was with God, and the Word was God. The summary of Jesus Christ is made clear in the end of John 1:1, as he emphatically declares that Jesus is God. If it was not clear to anyone in verse 1, then he states in verse 14 very clearly that the *"Word became flesh and dwelt among us."* There can be no mistake that John is saying that the God of Creation came to the earth that He created to inhabit human flesh. This is the same God that spoke the worlds into existence in Genesis, but in the New Testament robed up in the flesh of a little baby in Bethlehem.

As surely as the Bible teaches that there is only one God, which is often referred to as monotheistic, the Bible also teaches that there are three distinctive persons of the Godhead. It may be better to use the phrase Triune Godhead rather than Trinity to minimize confusion for those of the cultic persuasion. The God of creation has revealed Himself to mankind over the ages in three distinct persons. Each person of the Godhead has His own work to perform.

When one arrives in Heaven, he should not expect to find three Gods, but rather only one, who will be known as Jesus Christ the Savior of the world. Jesus is the clearest of all revelations that God has ever made known to man. The best illustration at this point to explain this phenomenon of three persons that are one would be a personal one. This writer is three different people, while still only one. My father, William Pratt, only knew me as his son, and that is the only relationship that he ever had to me. My son, Jason, only knows me as a father and this is the only way that he relates to me. My wife, Linda, only knows me as her husband or companion in life, since she is not related to me as father nor son. Thus, my relationship to each of the three differs and each requires a different role to perform.

It would be totally inappropriate for a son to relate to his father as if he were his father's father. Each person of the Godhead has His own distinct role, but yet they all three are One. Even though just as my role as father, son, and companion are different, all three are still rolled up into one being. This interesting phenomenon probably will never be satisfactorily explained to man, so it must be accepted by faith.

God identified Himself as a Father figure to the Old Testament saints, such as Adam, Noah, Abraham, Isaac, Jacob, and others. They each referred to Him as God the Father. In the New Testament, people saw Jesus the Son as the God of creation. Jesus should be referred to as the God-Man, since He was completely God and completely man. That means that He was 100 percent God and 100 percent man at the very same time. He had every quality of God within Himself, yet He also had every quality of man within at the same time. Jesus was not half God and half man, but rather the God-Man.

In his letter to the church at Colossae, Paul wrote, *"For by Him were all things created, that are in heaven, and that are in earth, visible and invisible, whether they be thrones, or dominions, or principalities, or powers: all things were created by Him, and for Him: and He is before all things, and by Him all things consist"* **(Col. 1:16–17**). Then it was Jesus who called the worlds into existence.

Jesus spoke of another that would come to minister in His place in **John 16:**7. When Jesus spoke of the Holy Spirit coming to minister in His stead to the disciples, He was speaking of the completeness of His earthly and fleshly mission. The God-Man had to leave in order for the Holy Spirit to come to complete His mission on earth.

The Holy Spirit is not prohibited by the fleshly body that housed Jesus the Son. God the Son could only go as far as His legs would carry Him in one day. That was usually fifteen to twenty miles in the Judean hill country, but for Jesus, it usually was much less. People approached Him all the time for ministry needs, which generally slowed His pace considerably. However, the Holy Spirit is not bound by time and space, since He is at every place at the very same time. There is only one place in all the world that the Holy Spirit is not at this very second. He is not inside the heart of all people at this moment, but He surely wants to be. So the Holy Spirit is the eternal God that moved over creation, but with a very specific function in the present world to convict people of their sin, of the righteousness of Christ, and the judgment of a Holy God **(John 16:8).** The Holy Spirit comes into the hearts of those people that invite Him into their lives as they repent of their sins. He takes up residency within every human heart that sincerely invites Him into their lives that asks for a pardon from their sins.

Joel (2:28) said that in the last days of the world that the Lord *"will pour out my Spirit on all people."* This prophecy came to pass on the day of Pentecost when the Holy Spirit entered the world after the earthly departure of Jesus. The third person of the Godhead is present today to convict and bring people to Jesus Christ the Savior of men. All three persons have a different role, yet all three are one. This is a great mystery.

God Has an Eternal Plan and Purpose

Perhaps, Isaiah reveals some wonderful insight about God's plan for the ages when he stated, *"For thus says the Lord, Who created the heavens, Who is God, Who formed the earth and made it, Who has established it, Who did not create it in vain, Who formed it to be inhabited: I am the Lord, and there is no other"* **(Isa. 45:18).** He had no intention of leaving the land *"in vain"* which means empty or in waste.

The Lord had a definite plan in creation for man to inhabit His world and that He would be the Ruler with no one else. Since He is the sovereign God who owns and rules the world, then he has the right to plan anything that He desires. God did not haphazardly put all things together, but rather had a marvelous, loving purpose for everything that He did in creation.

God did not create from a need for Himself, or a deficiency, but He created out of a life of self-sufficiency with plenty to share with man. It was His free will that brought creation to exist. God carefully planned every detail and not anything nor was anyone an afterthought or an evolving process. Every aspect of His creation was perfect leading Him to say *"it was good"* on five occasions and then finally saying *"it was very good"* in **Genesis 1:31** after God had created man.

In God's marvelous plan, He designed man to have dominion over the animals and all creation. God breathed into man and he became a living soul. Man does not have a soul, but is a soul. The body is mortal while the spirit is immortal. He intended for man to live in harmony with other men and the world around him. It only makes sense that if the great God of creation had such purpose in creation, that He surely has a plan for every man today.

Solomon who wrote the Proverbs stated that God has perfect wisdom of everything by declaring He had a perfect plan, *"before his works of old"* **(Prov. 8:22)**. *"From the beginning, before there was ever an earth"* and *"while He had not made the earth or the fields, or the primal dust of the world. When He prepared the heavens I (Wisdom) was there, when He drew a circle on the face of the deep, when He established the clouds above, when He strengthened the fountains of the deep, when He assigned to the sea its limit, so that the waters would not transgress His command, when He marked out the foundations of the earth, then I was beside Him as a master craftsman; and I was daily His delight, rejoicing always before Him, rejoicing in His inhabited world, and my delight was with the sons of men"* **(Prov. 8:23–31)**.

Solomon had been informed of God through his gift of special knowledge that it was God who had designed the earth and all of the intricate details of the planet. God set the boundaries for each element which makes the earth to function, such as the water above with a circle, the clouds, the seas with their limits and every detail; however, all of this was for His great delight in the sons of men. The idea that the greatest delight of God is in His creation of mankind, which is made in His own image. It is well documented that God is from everlasting to everlasting, but His most magnificent creation is that of man that will bring their Father great glory. The eternal God of the Ages desires to share His wealth of blessings with others for the fellowship of others.

Man is God's highest level of all that He created, which indicates that mankind is the most important creature that can receive some understanding of the eternal God of glory through revelations. God revealed Himself through the first man Adam by walking and talking with him in the Garden. He revealed Himself to Moses through a burning bush in the backside of the dessert and later on the tops of mountains. He spoke to Isaiah by telling him that a Son would be born in due time out of the Branch of David that would reign in a Kingdom with no end. The same God revealed to John the Baptist that Jesus was the Anointed Savior that he was unworthy to baptize, yet he obediently did. Those revelations are still coming to men and women as God is still calling out people to serve Him in proclaiming the gospel in mission fields around the world as Christian servants. Thus, God has a plan and a purpose for every created being that He has made. However, man must seek God for Him to make that revelation clear. The great prophet Jeremiah was given a message from God for all ages of people. God said to **Jeremiah 29:11**, *"I know the plans that I have for you, says the Lord, they are plans for good and not for disaster, to give you a future and a hope" (NLT).* It is apparent that God had a plan and purpose for every single thing of His creation as nothing came by happenstance nor chance.

God Has an Eternal Nature

Herein are a few of the frequently discussed facts concerning the nature of the true God of Eternity. Hundreds of good books have been written concerning each of these brief overviews regarding the nature of God. The following is merely some summary statements that have been gleaned from the Bible and studies by the author.

The true nature of God does not change, since He is 100 percent good and perfect. The prophet **Malachi** stated, *"For I the Lord do not change"* **(3:6)**. If He were to change, then He would have to be less than perfect and He refuses to change to an imperfect being. It is impossible for God to change, since **Hebrews** declares that He is the same yesterday, today and forever **(13:8)**. Also, the Lord has a divine love that is all excelling, which is 100 percent love without any degree of hate. This attribute makes it impossible for God to hate any person. Although God does hate man's sinful behavior, He never hates the sinner.

One of the simplest and yet most profound statements in all the Bible about the God of creation is made by the Apostle John, *"God is love"* **(1 John 4:8)**. John continues by stating that God first loved us **(1 John 4:19).** In reality, love is a decision and God had already decided that He would love every man before He spoke anyone into existence. He literally knew each person and what each one would do, yet His nature is still to love every person unconditionally.

God's nature is to love every man, no matter what he has done or is. That means that God died and suffered for every man that He loves, which is every man, woman and child.

God's perfect love is patient and longsuffering, which means that God is willing to wait for man to accept His love even if the Lord has to wait for years and years. His great desire is to receive man's love without coercion or manipulation. God's longing is to be the object of man's love and affection. This would be man's reasonable response, since the Lord provided all the needs of man in the garden that he would ever need or want. It should be man's natural response to such great love from his Creator.

God's perfect love will never change, because He never changes. The Bible declares that God is always dependable and will not change any part of His character. Even though God has perfect knowledge and perfect foreknowledge of things that will happen, He does not change His mind or plans. All things have been set into motion by this infinite Creator.

Another natural attribute of God that, perhaps, stands above all others is that He is absolutely holy. He is so pure and full of truth that there is no error in Him. He has never told a lie, and He cannot lie. He has never committed one sin and cannot sin. He is never jealous, partial, or unfair, thus He is holy in character. Even the angels of Heaven sing and praise Him for His righteous character by singing "Holy, Holy, Holy" **(Isa. 6:3).**

God's nature of holiness and righteousness differs greatly from that of man, which is borne with the nature of sin. While the flesh of man is drawn to commit sin and rebel against good, God's nature is just the opposite. He hates sin and rejects the sin in man's life. Those who truly *"love the Lord hate sin"* **(Ps. 97:10).** God cannot over look any sin, no matter how insignificant that man considers it, because of His holy character. The Psalmist declared God as *"The Holy One of Israel"* **(Ps. 78:41)**, and says that *"the Lord is righteous, He loves righteousness; His countenance beholds the upright"* **(Ps. 11:7)**.

God is perfect in holiness and that nature will never change, since God refuses to change. If the Lord were to change, then He would have to change to a position of less than His current state of absolute holiness and purity. Thus, He will not and cannot change His nature of purity. He is already 100 percent pure in holiness and righteousness and that cannot be improved upon. God has not grown any older, nor has He lost any of his strength or power. He will not change. Once again, the Bible says, *"He is the same yesterday, and today, and forever"* **(Heb. 13:8).**

God Is Eternally Sovereign

God has always been in charge of His creation, even though many disagree. According to God's Word, He simply brought things into existence from pure nothing by speaking. He had complete control and authority over all His works in creation. The term sovereign means that one is the supreme ruler over. Indeed, God is in complete control over everything and everyone as the Divine ruler. It does not take a genius to comprehend that one would not create something that he would not be able to control. God has perfect wisdom and perfect foresight that permitted Him to see all the way down the road ahead for thousands of years or even billions of years ahead.

God takes no directives from anyone and makes His own decisions without the counsel of anyone. Since God has never lost any of His power, then it is certainly true that He is totally in charge of the universes today, even though, it may not appear that that is the case. Remember that things are not always as they seem to appear.

A term that often describes the perfect Divine nature of the Creator is the word Omni, which simply explains the perfection of the Lord. Three perfect descriptions are ascribed to Him.

A. God is Omniscient or that He has perfect knowledge of everything that has happened or that will happen. Someone wrote a gospel song that says, "has it ever occurred to you that nothing ever occurred to God." Nothing that man can ever do will catch God off guard or by surprise, because He is completely knowledgeable of every move and every choice that man will ever make in his life. There is nothing that is hid from the Lord, since He knows all. In fact, John declares, *"God knows everything"* **(1 John 3:20).**

B. **God is Omnipotent** or that He has perfect power to do anything that He pleases. God is absolutely unlimited in what He can do and nothing is impossible for Him to perform. The Psalmist says, *"the voice of the Lord is powerful"* **(Ps. 29:4).** He has never lost one ounce of His power. God never gets tired and weary and loses one step of pep. When man toils and labors, he loses energy, but not God.

C. **God is Omnipresent** or that He is at every place at the very same time from one end of eternity to the other. There is only one place in all of creation that God is not at this second. He is not in the heart of the unsaved man and He will not enter until He receives an invitation from that lost man. No one can actually leave the presence of the Lord while living upon the earth. King David said, *"Where can I go from your Spirit? Or where shall I flee from your presence? If I ascend up into heaven, thou art there: if I make my bed in hell, behold, thou art there"* **(Ps. 139:7–8).** This attribute like the other two only belong to the one sovereign God of the universe.

Chapter 6

The Eternality of the Creator

1 Mohler, R. Albert, ***Decision Magazine***, Billy Graham Evangelistic Association, Charlotte, N. Carolina, December 2013, pg. 14–17.

2 W. O. Carver, PP (page 48).

CHAPTER 7

THE FIRST CREATION STORY

Genesis1:1–2:3

This is the historical record of a loving God revealing to His people how the world and all of life came into existence. According to the Bible, God brought all physical things, including the planet earth together approximately four thousand years before the coming of Jesus in the New Testament era. The teachings of creation seek to answer the most primary questions and many other puzzling questions of the mind of inquisitive man. The teachings in the Bible hold the keys to understanding the entire creation account, as well as, the origin of the human race. Nothing happened upon its own, as there is always an intelligent being behind every invention that has ever been created. The writer of **Hebrews 3:4**, states this great truth well, *"For every house is built by someone, but He who built all things is God."* Yes, there is a Supreme Being that set a very sophisticated plan into action and created everything that has been created. Perhaps, God designed a very complex creation, so as the most brilliant person could never discover the full extent of everything about the Creator nor a full detail of all His creation.

However, this same God wants man to know that He is the Master Architect or Master Engineer, which designed and created all things. He was willing to share his fabulous love and wealth in order to receive man's love and admiration in return. *"Know that the Lord, He is God: It is He who has made us, and not we ourselves"* **(Ps. 100:3)**. Then in **Psalm 104:3**, *"Who walks on the wings of the wind,"* so how can any man understand His creation without the Bible teachings? *As for God, His way is perfect; the word of the Lord is proven* **(Ps. 18:30).** Even though man will never be able to grasp the fullness of the Creator, every person will have the privilege of personally knowing the indescribable Creator of all.

The very first verse is truly the foundation for the whole Bible. If God truly created all things, then He controls all things and He can do all things **(1)**. This same verse also, refutes all of man's false philosophies concerning the origin of the world: It refutes atheism, since the universe was created by God; it refutes pantheism, as God is transcendent over His creation; it refutes polytheism, for one God created all things; it refutes materialism, for all matter had a beginning; it refutes dualism, because God alone created all; it refutes humanism, because God, not man is the ultimate reality of life; it refutes evolution, because God created all things **(2)**.

God's Word is always true and anything that contradicts the Word of God will fail to stand the tests of time. Most of the philosophies of man are arguments that something else or someone else brought forth creation, instead of God. It is an act to dethrone the God of creation from receiving the Glory of His work. This discussion will be centered upon that One God who designed and created all things according to the Genesis story. If one will accept this Bible story by faith, then he will have to reject the false stories that have arisen over thousands of years. It will be impossible to embrace any other version of creation with accuracy, if one accepts the biblical account. No person can accept both views nor seek to combine any two of these accounts as accurate. Each person must determine which story that they choose to accept as truth. It requires a belief in one version or the other.

The Evidence for Seven Real Days of Creation

First Day

One day, God the Spirit spoke all things into existence that have come to exist. The Psalmist said that God "*commanded and they were created*" (**Ps. 148:5**). He is referring to the heavens, mankind, the angels, sun, moon, stars and the waters. Matter is not eternal as some religions claim today, but rather all material things had a beginning, since God called them into existence. Before the account in Genesis 1, nothing physical existed, since all things were spirit in nature. The writer of Hebrews states, *"by faith we understand that the worlds were framed by the WORD of God, so that the things which are seen were not made of things which are seen not made of things which are visible."* **(Heb. 11:13).**

In **Psalm 102:25–26**, the Psalmist declared, *"Of old You laid the foundation of the earth, and the heavens are the work of Your hands. They will perish, but You will endure; Yes, all of them will grow old like a garment; like a cloak You will change them, and they will be changed."* Even that which seems to be permanent and fixed will pass away and melt with a fervent heat as this old world will be gone. The "same" **(v. 27)** Creator will make a new place for all people that choose the Creator, which will "have no end." The writer of the Psalms is making a prophetic statement about the Creator.

"Everything that was made was made by God; and without Him was not anything made that was made" **(John 1:3).** He simply is the Author, Creator, Builder, and Sustainer of the entire universe. It was all for the pure pleasure and providence of a loving God, who desired to share His fabulous wealth with creatures designed to act like Himself or *"made in His image."* The new physical things would come to know their Creator, without actually seeing Him in the beginning. He merely spoke to bring all physical production into existence. **Nehemiah** declared, "*You alone are the Lord; You have made heaven, the heaven of heavens, with all their host, the earth and all things on it, the seas and all that is in them, and You preserve them all"*

(9:6). God not only made all the physical things, but He preserves all things that He creates.

What is the Genesis writer's thought when he uses the phrase or statement, *"In beginning"* of that very first verse? The word in Hebrew (bereshith) means "origin," "source," "generation," "beginning," or "starting point." Could the story be centered upon the beginning or the origin of God? Absolutely not, since God had no beginning, because He is an Eternal Being. If "the beginning" does not refer to God, then it must refer to the beginning of man and the universe that man inhabits. Yes, this starting point refers to an absolute beginning of mankind. The word Genesis means "source or origin" of something other than God, who is the source or origin of all things. Unfortunately, much of the world's teachings today deny any absolutes, such as moral absolutes; however, the Bible declares that it is a Book about absolutes as there are many absolutes in this creation. The world seeks to make man less responsible by claiming there are no absolutes, but this world had an absolute beginning with an absolute Creator. The Bible leaves absolutely nothing up to chance.

The writer states that in the beginning of the world, there was "God," which is the same eternal Being that **John (1:1)** called the "Word" in his message. The Hebrew word for "God" is Elohim, which is derived from a root found in the Arabic *meaning "to fear" or "to reverence"* **(3).** That word "Elohim" occurs 32 times in the first 31 verses of the Bible along with 11 personal pronouns referring to God. It is interesting to note that the Bible commands man to fear only God and not man nor the devil. This is not the type of fear of terror, but rather it means to honor and respect the Creator of life. The Sovereign God who controls all things commands man to respect His Divine Being. The term "God" is used 2,570 times in the Old Testament **(4).** God (Elohim) is the same One that told Moses that "I AM," and is the same God that John in his Gospel presented in 7 cases as the great "I AM." John declared Him in the following ways: "I AM the Light of the World", "I AM the Living Water," "I AM the Bread of Life," "I AM the Good Shepherd," "I AM the Resurrection of Life," "I AM the Way, the Truth and the Life," and "I AM the Lamb of God."

The creation accounts describe the action of God as "created," which is a Hebrew word (bara) that is only used as an activity of God. This Hebrew word is never used in connection nor reference to any work of man. "Created" is used as a verb three times in the first creation story. It is used in **Genesis 1:1**, as God created matter or something physical. The second usage is found in **Genesis 1:21**, as He created beings that have life such as the sea creatures. Then, the word is used as God created spiritual life in **Genesis 1:27**, when He created man in His own image. The literal meaning of created (bara) means "to create out of nothing."

Thus, it has been said that God came out of nowhere, stood upon nothing and called a physical world into existence without any witnesses, except God alone. This story features not one single eye witness other than the sole Creator. The question is, "Why did God not permit anyone to witness the happenings of the beginning of civilization, so that someone could have written all this down or reported on such an important happening?" God could have created man first and then allowed him to see all of the happenings from the start. All powerful God could have done it that way, had He chosen to do so.

One must believe that God did what He did for a good reason by bringing man onto the scene as the crowning work of His creation. Image the Creator God as a highly experienced chief executive officer (CEO) of an incredibly large company with millions of employees simply laying out His plans of creation one step at a time. In this case, there were no employees, with just the CEO speaking into existence what He desired. The Creator simply used voice commands to bring all things into existence from nothing. There are those that would have trouble believing such an account of creation. However, Jesus, the God Man used those same voice commands when he spoke to a dead man in a cemetery called Lazarus to come forth from the grave **(Jn. 11:43**) and before His disciples in the midst of a serious storm in a boat on the Sea of Galilee, when Jesus said to the storm, "Peace, be still" **(Mark 4:39).** As quickly as the Creator gave these voice commands, all things, whether spirit or physical quickly obeyed and responded completely to the voice of the Creator.

Since the Bible teaches a special element of belief in the unseen, which is called faith that permits man to see without seeing any fleshly substance. Man must come to the realization as he looks around at creation by seeing that creation did not just happen by itself or by chance of things aligning up in certain positions to bring it into existence. The Bible teaches man can accept the biblical account or some other method of creation according to his own belief system. God created man with the power to reason and choose, since the nature of God was placed into man and no other created being possesses this spiritual insight. This belief in God must become real by each one's personal faith in a Creator that is Supernatural, yet He can be seen without a fleshly eye. Spiritual sight is far more advanced than the physical eye could ever envision. Jesus told His disciples one day that could not see the Spiritual, *"Seeing they may not see, and hearing they may not understand"* **(Lk. 8:10).**

From the Hebrew expression, coming from nothing, and "*Ex Nihilo*" *(Latin)* means that The Great Creator manufactured the worlds from absolute nothing physical. In one moment, nothing existed; however, immediately after God spoke, real physical substance came into existence. There has neither been one product made by human beings nor an animal that has ever been able to make something from nothing. Only God can take nothing and perform such an incredible feat. Man is capable of taking lumber from a tree to build a house, but he cannot build a house without materials. Thus, God made time as well as, space from nothing, since there was not even thin air prior to His speaking all things into existence.

God gave a voice command for each of His works of creation and things immediately heeded His voice authority beyond any other created being. In **Psalm 29**, the Psalmist, David declares a perfect seven times, *"The voice of the Lord is over the waters, is powerful, is full of majesty, breaks the cedars, divides the flames of fire, shakes the wilderness and makes the deer to give birth."* Then in **Psalm 33:6**, the writer stated, "*By the word of the Lord, the heavens were made*" and then, "*for He spoke, and it was done*" **(33:9).** Again, the Psalmist declared, "*He uttered his voice, the earth melted*" **(46:6).**

The next phrase and the object of God's work is *"the heavens and earth."* The word ending for *"heavens"* is plural which indicates more than one, while the earth is always singular. The heavens and the earth include some one hundred billion galaxies, each with some one hundred billion stars **(5).** If scientific facts are even half correct, then the heavens and earth that man knows about is only a speck on the horizon of all of God's creation. Dr. Edward B. Lindaman suggests for man to think of the sun as being the same size as the period at the end of this sentence and if it were scaled down to the same degree, our sun's nearest stellar neighbor would be another dot about ten miles away **(6).**

While the heavens are mentioned, this unfolding drama centers upon the production of the earth rather than any of the other planets. All these other planets were created, but the author wants to specifically talk about those events which happened on planet earth. The second verse begins with the birthing of the earth as the writer states that it was *"without form and void."* While the time period in the first two verses is unknown, some scholars see a huge expanse of time that could be billions of years. Their assertion is that the earth just laid vacant for these billions of years, while others contend that there was another earth that included pre-historic creatures. It does not seem correct to think that God began the production of creation of the planets and then ceasing the project for millions of years. The God that created all things had a definite plan of action, since He knows all things, plus He possessed all the resources to complete the task. That view does not seem to fit with the God of creation that desires fellowship with man which was made in God's image to find fulfillment.

The further claim is that time is of no essence to God, since a thousand years to Him is like a mere day to man. While there is some truth to this logic of man, God is certainly interested in time, since He says for man to redeem the time **(Eph. 5:16).** Many of this persuasion refer to the time lapse as the "Big Gap" theory, which happened after verse 1. Those that support such a view of Scripture believe that science correctly reports the age of the earth in billions of years, so this removes any scientific conflict with time. The biggest problem with this popular view is that one assumes the scientific speculation of years and then attempts to reconcile the Bible record to the theory. The truth is that the findings of the science world continue to be in constant change in their discoveries. Science is always searching for truth, while the Bible is the book filled with only truth. The very nature of science is to find out how things were manufactured or how they continue to exist. If there is a conflict between science and the Bible, then the Bible is always correct, even though the Bible is not a book of science.

The Believer is given a tremendous insight as the writer uses the words *"without form and void"* (tohu and bohu). These words simply suggest that the earth was in an incomplete stage of production. They further declared that the final shape had not yet come and that the earth was empty. It would be like a new house under construction that had merely been framed with walls, ceilings, a roof, but contained no furnishings with no occupants at that moment. God had not yet finished His work of creation. He simply was slowly preparing for the entire picture like a trained artist painting a complete portrait. God had multiple strokes to make upon the canvas, developing a certain production each day. He was setting the stage for the most magnificent component of His creation with a dramatic climax. Absolutely, nothing would surpass the last scene when He finished His painting with man in the very center of the garden of paradise. Man would be manufactured in the "image of God."

Another interesting part of the development was that "darkness" reigned upon the earth and overshadowed all of God's creation on day 1. A Believer might say, how can that be, if Jesus Christ is the light of the world? Some view this phenomenon as the presence of the devil after he was cast out of heaven upon the earth. Since Satan is the prince and ruler of darkness, his presence shrouded the earth. This was no ordinary darkness, since it was a "deep" covering over the earth's surface. For a follower of Jesus, it is totally inconceivable that Satan would have been within a million miles of the Lord during His deliberate works of creation. That would mean that if Satan was overshadowing the Master that he would have more power than his Maker. The truth is that this being had not yet been created. See the teachings on the evil one in chapter 5 of The Pre-Creation Story under the section of "Who is Satan?"

In light of the word translated as "waters" at the end of the statement, scholars have concluded that Satan's very presence perverted God's work. Some interpreters claim that this was the evidence of another earth before the present earth that was destroyed. However, this does not coincide with the description of John in the story in **Revelation (21:1),** when he described this as the first earth and not the second.

The word for "waters" means stormy melted waters, which would strongly indicate a prior frozen state. This phenomenon could help to explain signs of an ice age in some findings in many parts of the world. The word "deep" refers to a place that was devoid of any heat. Thus, the writer is describing an earth that could have literally been covered with a sphere of ice which was in a very dense state of cold darkness. Some have suggested that this was the first ice age that covered the face of the earth, which was caused by Satan's presence upon the earth. Those that believe this theory say that Satan had just experienced his first fall as described in **Isaiah 14.** This argument possess a lack of merit in many ways and is discussed in chapter 5 of the Pre-Creation presentation.

It is hard to image Satan arriving prior to Adam and Eve, since sin had not entered into God's creation. He indeed led an open rebellion in heaven and was cast down to earth with his followers, but that happens later. Perhaps, the Psalmist **(Ps. 46:6)** explains this point of the frozen dark matter, when he states, "*He uttered his voice and the earth melted.*" Although Satan was cast down to earth later, so it does not indicate that he actually ruled the incomplete earth for millions or billions of years. God lived alone prior to Genesis 1, so no devil or evil existed.

All this sequence of events from Satan's fall and the progression of creation did not happen as some suggest in a simultaneous event with years of delay, such as the Big Gap supporters suggest. Even though, God is patient, there is no indication in the Bible that he ceased His work to give Satan a field day so that evil could catch up with good. Some scholars contend that while Satan was in complete dominion upon the incomplete earth, there was chaos. It hardly seems fitting for the Lord to allow such adversity to His sovereignty, since He is the God of order. Keep in mind the Master Artist slowly developing a masterful work of art with continual strokes on the canvas seems to better fit the narrative. Once God started this project, there seems to be no cessation, but rather a deliberate continuous activity of creation.

"*The spirit of God*" was moving over this deep and dark sphere, which quickly changed the entire situation. Notice the reference to the Holy Spirit of God. John declares that "God is a Spirit" Being **(John 4:24).** The idea is that of God quietly brooding or admiring His work of creation. However, it was God doing His work and most likely Satan did not come to earth until possibly years later in **Genesis** chapter 3. Quickly all things changed as God spoke and gave His voice command, since things immediately respond to God's command **(v. 3).**

Things happened, when God began to speak, because His word is sharp and powerful like a two-edged sword. The sword sliced through the thick darkness as if it did not exist and brought the shining light of glory upon the horizon. Jesus declared Himself to *be "the Light of the world*" **(John 8:12).** Just as a new day springs forth with light, so does the first day of God's creation dawn by giving birth to light. God knew that everything in His world would depend upon light for life, growth and substance. Light brings warmth to the seed planted in the soil that is critical for germination. The Creator was putting together an intricate, super complex system that sustained everything, including man. When new seedlings spring forth they depend upon light to energize them and to be their power source for the plant process called photosynthesis, which involves the making of plant foods. That process totally depends upon light from sources such as the sun or some artificial light source.

From the discoveries in the realm of plant science has validated the work of God from the beginning of time. Without the ongoing process of food production in plants, there would be no oxygen for man and the animals to breathe. It would be a futile and fruitless life without light, because no child could ever grow and mature to adulthood apart from light. God's ecology is intricately woven together with each phase depending upon all others. Thus, the design of the universe and the wonderful balance within demonstrates a master architect behind the scenes planning each event and then calling the shots. It becomes a mind boggling puzzle for the person who honestly studies the Divine system to see how everything functions in harmony depending upon the *"Light of the World."*

The heavens and earth are so intricately connected, as well as, the other planets of the solar system all working in harmony with each other. Some of these discoveries became widely known through a God fearing scientist, Johannes Kepler in his discovery of the invisible forces of the moon pulling impact upon the ocean tides. Kepler, a German Protestant was born in 1596, was a follower of Jesus who saw a connection in all that God created. He believed that all of creation depended upon each other part of creation. Other scientists, such as Sir Isaac Newton identified the magnetic force found in universal gravitation between the moon and ocean tides.

Truly, one could say that the God of creation came into the matter that he created bringing the light and energy that would sustain all life upon earth. According to the Bible, that light came immediately upon earth in obedience to the commandment of the Lord—*"Let there be light"* **(v. 3).** God did not toil nor labor, but merely spoke things into existence because His word is law. The **Psalmist (33:9)** said, *"For He spoke, and it was done; he commanded, and it stood fast."* The light was turned on by God's face on the earth since the sun had not been created. No other being has such power. The writer John reveals the answer to this question of "where is the source of this light?" **(1 Jn. 1:5)** John says, *"God is light and in Him is no darkness at all."* Very often in the Bible, especially in the Gospels, the word light is associated with sight, particularly spiritual sight. The Bible declares that the voice command of God is all that is needful for things to come into existence, since His commands are executive orders. Ten times, the writer records the same executive commands are given in Genesis chapter 1, "And God said," thus it happened.

Someone has stated that these are God's "Ten Commandments" that have never been broken, unlike the other ten from Exodus 20:

that have all been broken. The **Psalmist (29:4)** said, *"The voice of the Lord is powerful; the voice of the Lord is full of majesty."* When God spoke, light immediately flashed across the scene of creation for the very first time. It had to be the light of His Divine Word, since the sun, moon nor stars had not been created on day 1. None of these appeared until day 4.

In heaven, there is no need for the sun, moon nor stars, since *"The Light of Glory or the Lamb of God"* will be present **(Rev. 21:23).** Like a master artist, God looked at His work and declared that *"it was good"* ***(v. 4).*** This is the first of six such statements that God makes about His work and communicates His satisfaction with what He had designed and created.

Then the Lord created the very first day by dividing it into light and darkness. This explanation gives the Believer the insight that God knew ahead of time that man would need rest every day. Man would not be able to function twenty-four hours at a time without a period of renewal. So in preparation for man's arrival, God designed a separation of two distinct parts of every day. One part of each day would be designed for activity and the other for rest and renewal. The actual commandment of the Lord is not recorded, but the results are. The light gathered into one end on that first day and the darkness into the other portion. One big question about the word "day" or "yom" in Hebrew was this a period of time or was it 24 hour period. It seems obvious to a student of God's Word that this is definitely a twenty-four-hour time period.

Each of these six days seem to be the first actual twenty-four-hour days on earth and there is no reason for the Lord to have changed any of that. Each day has two parts and it always begins with the evening portion before the morning portion in Jewish thought. This same pattern is found elsewhere in the Bible, when a day is mentioned. There is no reason to see each of these days of God's creation in any other form, other than the way it was intended as a twenty-four-hour time period. The story of Jonah in the belly of the fish was 3 days and nights as the same Hebrew word is used to explain the period of time. When evolutionists make their case, it has to be a period of years, usually millions of years. Why should

Believers have to accept their plan and give credence to anything that helps them deny God's story or God's glory. Nowhere does the Bible ever say millions or billions of years exist.

Then the Lord named each part of His first day **(v. 5).** The lighted part "Day" and the dark part, He called "Night." These two parts became the very first day of God's work of creation.

Today, there are debates over the meaning of the word "Day," as to whether this was a twenty-four hour period of time, or whether it was being used in some other measure of time. The Bible actually uses the word (yom) in three different manners. First, the term is used to describe a period of time that is very brief and much less than twenty-four hours when it speaks of "the day of salvation" **(Acts 16:31),** "the day of Christ's return" **(1 Cor. 15:52),** or "the day of God's wrath in judgment" **(Rom. 2:5).** The second usage of the term "day" describes a period longer than twenty-four hours. It is the day of eternity **(Eph. 4:30),** or the Tribulation **(1 Thess. 5:2).** Thirdly, the term refers to a twenty-four-hour period of time, which is a literal event in the creation story. The Lord clearly sets the time and seasons into motion and the planets are rotating around the sun in day 4.

Nothing afterward ever changed this constant system of rotation which teaches a literal twenty-four hour period of time. Some contend that the day was used in a symbolic manner; however, when symbolism is used the literal must always precede the symbolic. Thus, this is a clear statement of truth in Genesis and not symbolism. In **Exodus 20:11**, the Hebrew writer says, *"In six days the Lord made heaven and earth, the sea, and all that in them is, and rested the seventh day."* Here is a clear statement in the Law of God that defines the period as a literal week.

It often becomes curious to the student of God's Word why that evening proceeds morning on the first day. The writer states *"the evening and morning were the first day."* By the time the writing was recorded, hundreds of years had elapsed and Hebrew thought had been well advanced. God permitted the writer to express that lifestyle of Judaism, where each new day began at six o'clock in the afternoon.

Second Day

The Bible records God's second commandment ***(v. 6)*** *"Let there be a firmament."* This is not an easy word to translate (raqija), but can also be translated "expanse," "curtain," something "hammered out" or "stamped out." The idea is that of completing the work on a substance already under construction as a potter shaping, molding and fashioning the clay into the final shape. Since God did not need to toil nor to hammer out the final product, He simply spoke and His powerful word did the fashioning of the product. The matter that God created in verse one, likely had little resemblance to the sphere that God formed in day 2. The statement seems to confirm the fact that the "firmament" and "heaven" are essentially synonymous terms, both meaning "space" either space in general or a particular region of space, depending on the context **(7)**.

The expanse or firmament would include the waters below and the waters above the earth ***(v. 7)***. Some translate this solid matter as sky that was filled with water. Actually, the waters were separated out from the solid material, leaving a belief in the "canopy theory" of water surrounding the earth. This idea is that of a water vapor barrier that completely encompassed the sphere of the earth and protected all the earth from harmful radiation from the sun. This water vapor barrier was suspended in the upper atmosphere around the earth. The surface of the earth contained much less water before the flood than afterwards. The protective barrier was broken up during the flood as an act of God's judgment upon sinful man. According to the Bible, the consequences of sin not only impacted mankind, but also the entire planets that God had created for man.

There is much physical evidence to support the idea that the water under the earth indicated that there was a significant amount of water trapped inside the belly of the earth as it was being shaped into the sphere that currently exists. This would explain the presence of volcanic activity which spews out steam during eruptions. Most of the pressure from active volcanoes is simply steam from hot liquids. This concept is further supported by the breaking up of the great deep during the flood when massive amounts of water spewed out of the earth during the forty days of flooding. One can actually visit some of these sites where the hot steam rises from the earth in some of the Hawaiian Islands like Kona in a state park. This author visited this very site in 2019 to experience the powerful heat rising up from the rock basins. It produces such heat that a person may think that these are outdoor furnaces.

God clearly "divided" **(vs.** 7) the waters into those that hang over the firmament that formed the atmosphere in the heavens and those that were inside the sphere of the earth. Additionally, God separated out those waters that formed the water vapor canopy from those that were on the surface of the earth. Thus, the hammered out atmosphere was suspended between the waters above and those below that were literally hung on nothing, which is consistent with **Job (26:7)** that says, *"He hangs the earth on nothing."* The only reasonable explanation for this phenomena is that the earth is being kept in the suspended state by the One who hung it. The following verse further explains the canopy cloud theory when it declares that *"God binds up the waters in His thick clouds; yet the clouds are not broken under it"* **(Job 26:8)**. The Psalmist further stated that *"the heavens declare the glory of God; and the firmament shows His handiwork"* **(Ps. 19:1).** Everything about the planet hanging in outer space points to a Divine Creator behind the creation and the sustaining of that creation. In order for creation to reflect the marvelous work of God, it had to be fashioned in a physical manner that would be spectacular to the very sight of man. Thus, the firmament was a solid expanse of sky that held the waters above.

These thick vapor clouds held trillions and trillions of tons of water that encircled the earth's sphere and were supported from a solid expanse of sky. The upper atmosphere water vapor remained intact to protect all life on earth from harmful sun rays while producing a pleasant sub-tropical climate all around the earth. Scientist, Dr. Joseph Dillow has calculated that the water vapor cloud that surrounded the earth was about forty miles thick, which would be sufficient to generate forty days and forty nights of continual torrential rainfall; whereas if these waters above had been water clouds, then the moisture in the current atmosphere, if precipitated to earth as rainfall, would be only the equivalent of less than two inches of liquid water-hardly enough to sustain forty days and forty nights of rainfall at the time of the Flood **(8).**

Some scientists have estimated that the water vapor that was suspended above the earth weighed approximately 54.5 trillion tons. Since water is 773 times heavier than air, what force or One is keeping the water suspended above the earth? Only the power of God can keep all such things in order. The constant supply of water held above the earth is maintained through the process of evaporation. That water vapor canopy was broken during the flood and the torrents of rainfall came down upon the earth, creating massive erosion, which helped to reshape the entire earth's surface. For example, according to John Baumgardner, the world's preeminent expert in the design of computer models for geophysical convection ***US News and World Report*** contends that a flood that produces a one-hundred-mile per hour runoff would have been sufficient to create the Grand Canyon and other massive geological features and to deposit the various sedimentary layers in about one week **(9).**

Baumgardner, who received a master's from Princeton University in electrical engineering and a PhD from UCLA in geophysics developed a computer program named Terra that demonstrates the process by which the earth creates volcanoes, earthquakes and the movement of continental plates. He believes that the earth is less than ten thousand years old and that a universal flood reshaped the earth's surface. His high tech computer program demonstrates a 150-day period of flood surging that resulted in a rapid retreat of waters as the continents began to emerge sending the global flood waters back to the oceans at speeds in excess of one hundred miles per hour. This model has produced an amazing picture of radical changes in the geography of the earth. This discovery has proven to millions that the Great Flood in the day of Noah was worldwide and not a local Middle East flood.

The bodies of water from rivers, lakes, and the seas prior to the flood were a mere fraction of the water left on earth following the flood. The earth's surface is currently made up of about 70 percent water; however before the flood it was likely less than a third water.

The atmospheric water vapor in the firmament that once protected man, animals, and plant life, now resides upon the surface of the earth. Since water helps to purify the air, there is a need for the water to exceed the amount of earth's land mass **(10).**

Prior to the flood vegetation grew at both poles, which explains the trapped traces of green vegetation remains found by scientists in the polar regions today. Much of the past history of the regions has been preserved in the vast deep freeze of these frigid regions. All parts of the earth were inhabitable before the flood. Even though man may not have traveled around the globe, the chances are great that the animal kingdom migrated throughout the entire earth. This is supported by the fact that bones and animal remains have been found in the frozen Polar Regions. Some fossils and sedimentary materials have been discovered within the frozen layers of debris in the Polar Regions, in rock and silt sediments elsewhere around the world. Also, it should be noted that bones and carcasses of man and beast were carried around the globe, even to the polar regions during the period near the end of the flood as the waters rushed around the globe.

The firmament is an indescribable complex work of God that man simply cannot totally explain. The writer declares that all this happened at the command of the Lord and validated the powerful spoken Word of God by declaring, *"and it was so"* **(v. 7).**

God once again named His creation on day 2 just as He had done on day 1 and called it "Heaven." The term "heaven" is frequently used word in the both testaments. God said to **Job (41:11),** *"Everything under heaven is mine,"* while the Psalmist **(Ps. 8:3)** declared to God, "*I consider your heavens the work of your fingers.*" Elsewhere, the Psalmist **(Ps. 115:15–16),** reported to the Lord, *"He made heaven and earth. The heaven, even the heavens are the Lord."* This statement would indicate the plural of heaven with more than one place called heaven. This would include all that was seen, plus the part that could not be seen by man physical eyes.

Paul stated that he *"was caught up into the third heaven"* **(2 Cor. 12:2)** If there is a third heaven, then there must be a first and second heaven. Some theologians explain the first heaven as the visible part surrounding the earth where the birds fly through. The Genesis

(1:20) writer stated, *"the fowl may fly above the earth in the open firmament of heaven."* The second heaven would be the sphere beyond the first that is invisible to the eye of man on earth or telescope and may be the dwelling place of the angels and all saints of the Lord that He will bring back with Him on His return mission. The third heaven would be the very throne room of God or the Holy of Holies, where no man is currently permitted to enter.

The Psalmist **(Ps. 103:19)** declared, *"The Lord has established His throne in heaven."* That first heaven included the paradise of Eden from the beginning and prior to the fall of man into sin. Truly, all of God's work of earth was Heavenly from one end to the other as He declared it good. This was the second day. The same term as day 1 is repeated of the evening and morning composed another day. Another period of twenty-four hours had passed and another stage of creation was completed.

Third Day

On the third day, God again gave a voice command that brought forth action when He commanded that the waters should be gathered into their places **(v. 9).** He now formed the dry land from the bodies of water. According to the biblical evidence in **Job (26:10)**, God drew the waters together and established boundaries for the place of the waters. Elsewhere, the biblical record declares that a decree was given to the waters and that even the waters obey His every command **(Prov. 8:29).** It is no wonder that the disciples of Jesus were so amazed when He commanded the winds and the seas to obey His voice and they did so immediately **(Mt. 8:26–27).** The waters and all of nature recognizes God and gives heed to His every command. Nature, unlike the nature of man, always obeys the Lord. When the Lord spoke *"it was so"* **(v. 9)** An appropriate action followed every directive that God spoke, because things in nature obey their Master. This is the second time that the writer uses this expression of obedience.

"God called the dry land Earth" **(v. 10)**. Again the Lord names the creation of the third day as He has in the previous days and calls it "earth," which is another term frequently used in Scriptures. This time God gives a proper noun to this portion of creation as He names it Earth. It often is coupled with heaven, which is a simple reference to all of God's created order. The Psalmist **(Ps. 95:5)** declared, *"And His hands formed the dry land."* The writer is merely expressing the activity of God, who is a Spirit Being in fleshly descriptions for the sake of His audience. These type of references are anthropomorphical (the physical character) terms that give human understanding to heavenly action. Already, it has been clearly stated that God merely spoke and His Word acted to perform every desire of His heart. *"By the word of the Lord the heavens were made, and all the host of them by the breath of His mouth"* **(Ps. 33:6).** Again, various writers use human terms in describing the work of the Divine Creator, since man comprehends the physical qualities much more easily than the spiritual.

The *"earth"* would become the domain for man, while the seas would be the habitat for water loving creatures. Many waters were gathered together to form a body that the Lord called the *"seas"* **(v. 10).** The Psalmist declared that the seas were the playground for large sea creatures as well as, the home for adventurous sailors. *"This great and wide sea, in which are innumerable teeming billions things of aquatic creatures, living things both small and great. There the ships sail about; And there Leviathan which you have made to play there"* **(Ps. 104:25–26).** The Lord concludes day 3 by repeating the expression that he used after creating light on day 1 by saying that *"it was good."* Of course, the very nature of God would keep Him from making nothing but good products of creation.

The Lord begins the second part of day 3 by again commanding action from His spoken word as He says, "*Let the earth bring forth grass, the herb that yields seed, and the fruit tree that yields fruit according to its kind, whose seed is in itself, on the earth"* **(v. 11).** Immediately this command was obeyed by the earth and all was performed, *"And it was so."* This is the third time that this phrase is recorded to allow the reader to understand how that the earth quickly obeyed the voice of the Lord.

God literally landscaped the earth and put a covering of vegetation over the entire surface. The landscaping consisted of a grass sod, fruit trees, herbs and all kinds of plants and vegetation. God not only provided the food for man on day 3, but He brought forth planting for an aesthetical value that would make life more pleasant and attractive for all creatures, especially man. The plant population became full grown instantly and produced fruit, seed, nuts, berries and other food stuff ready for harvest for God's first family upon their arrival. In fact, there would be a great need of vegetation for food for all the animals to feed upon when they arrived in just a few short days. God was making preparations and provisions for every creative that He had planned to populate His new earth. There would not be a single need when the creatures arrived. This certainly demonstrates how carefully every phase of creation had been considered very well in advance.

When a student studies the subject of botany, he can appreciate all the intricate structure of the plants, including the microscopic cell structure and realize that a Master Designer was behind the construction of every single plant and blade of grass.

Once again, the writer declares that the command of God was matched with an immediate production by exclaiming *"it was so"* **(v. 11).** The writer was describing an immediate multiplication of each type of plant that took place until all the earth was covered with an array of beauty and splendor. Each variety had produced and reproduced throughout all the earth and when all was complete, the third day came to a close. The same statement from the previous two days is made with God declaring, *"It was good"* **(v. 12).** Day 3 ends with *"the evening and the morning were the third day"* **(v. 13).**

Fourth Day

The new day again begins with a voice command from the Creator, *"Let there be lights in the firmament of the heaven to divide the day from the night"* **(v. 14).** This verse is definitely related to the previous description in verses six through eight where the Lord hammered out a solid production of sky called a firmament with a mere voice command. The Lord subdivided the previous production by hanging lights in each part of it. These new lights would provide several compliments to creation. The two sets of lights would governor over the days, years and seasons upon the earth, as well as, bear signs to the inhabitants to guide and direct them.

There are thousands of testimonies where lost or misguided people have looked to the lights within the firmament to guide them to safety. The Lord prepared even the heavens to be a guide and filled with signs to point people out of darkness and assist those in getting on the right course. The placement of the lights in the heavens were not haphazardly thrown together, but would hold definite positions, further displaying the hand of God upon them. The sun is a solar powered planet unlike anything that God called into existence. The sun's surface converts 4 million tons of material into energy, which is the equivalent to 10 billion nuclear bombs **(11).** Some of the sun's energy supplies the planet earth, which keeps the temperatures consistent in each of the four seasons. The sun is approximately 91 million miles from the earth in January and approximately 95 million away in July **(12).** The reason that the earth is actually warmer in July has to do with the 23.5 degrees tilt of the earth to the sun **(13).**

That tilt of the earth upon its axis most likely occurred during the Great Flood according to many modern-day scientists. Great pressure was exerted upon the entire planet with the new weights of water that had previously held in the water vapor canopy in the upper atmosphere. The weight of the new mega tons causing many low elevations to become lower, while the mountain ranges in many locations around the globe were being squeezed upward. It was during or following that tension upon the sphere of the earth that the entire earth tilted over by the 23.5 degrees that is found today in its current location to the sun.

The seasons were established on day 4 as the Lord added His touch of variety to the climate of the universe. There has not been any evidence that this system has ever changed or been severely altered in any manner. Often the seasons are cooler, warmer, more rainfall, more dry than the previous ones; however, the seasons are generally predictable. Frequently, those environmental extremists often declare that global warming is occurring in the universe, every time there is a summer season with elevated temperatures. However, they go in their closets during those extreme cold, harsh winters, but suddenly appear in the hot weather. The fact is that since records on weather have been kept, the weather and temperatures have fluctuated up and down during the same seasons. No season is identical to any previous seasons because of these continual fluctuations.

The seasonal changes help to create growing conditions for many crops, keeping migratory animals moving geographically, as well as, adding a variety to man's life. These are reminders that God is faithful, just as the seasons have never failed to come and go. The seasons, the days and the years will continue to come until the God over them decides that they have served their purpose and He brings them to a close. Until that hour and that day of which no man knows, we can count upon His faithfulness and control over these lights.

The writer states that the purpose of the lights are to give light to the earth **(v. 15).** Again, the voice command of the Lord says, *"Let them be"* is put into action with the fourth statement of *"it was so"* **(v. 14)**. It became a done deal as God spoke these elements into His firmament of heaven.

The Lord made the greater light (sun) to rule the day and the lesser light (moon) to rule the night. From the early days of creation, there was plenty evidence that men worshipped the sun, moon and the stars, as well as, other planets of the universe. Maybe the writer purposefully chose not to confuse these two great lights as objects of worship, but the One that made them. In the New Testament era, some men associated Barnabas and Paul at Lystra as the gods of Mercury and Jupiter **(Acts 14:12).** The writer may have purposefully used the terms of greater and lesser lights, rather than mentioning the sun, moon, and stars, so not to be misunderstood by some who already worshipped the lights.

The moon was once thought to be larger than the sun; however, modern science has proven that the sun is over one hundred times larger than the moon. The moon is about one-fourth the diameter of the earth, while the sun is over one hundred times larger than the moon. Even though the moon is less than a quarter million miles away, these two planets perform together for the benefit of the planet earth **(14).** It appears that all of the planets within the solar system with the earth work in conjunction with each other, which is a credit to the Creator, Genius that brought all things into existence. Without the moon, all clear nights in areas without artificial lighting would be too dark for many nocturnal animals, such as cats, dogs, mink, bats, owls, rabbits, coyotes to search for food **(15)**. That same Creator is sustaining those same structures that He originally called into being on day 4.

Notice that Moses was careful to state the two lights were the greater and lesser. He did not say the greatest light, since it only has been recently determined that there are stars that are millions of times larger than planets in earth's solar system. For example, the star Antares is 64 million times the size of the sun and several stars in other constellations are said to be even larger. Only the Holy Spirit could have guided the writer's thoughts and revealed such great truths about science. The universe is a vast display of the handiwork of an all-powerful God who has no limitations.

The statement seems so simplistic, *"He made the stars also"* **(v. 16),** yet consider the innumerable supply of stars that exist today. Scientists have estimated that there are over one billion stars in this galaxy with the Milky Way and that there are more than 200 billion other galaxies, which would total over 100,000,000,000,000,000,000,000 stars **(16).** The universe or the universes as the scientists of our world contend are mind boggling and huge in comparison to the earth. Even the creation of the billions of stars is recounted from an earthly perspective, so that the sun and moon appear as two great lights, while vast galaxies of stars appear as lesser luminaries, being mentioned almost as a footnote in verse 16: *"the stars also*" **(17).**

The Creator carefully planned all of these moving luminaries with boundless energy constantly moving and twinkling in the night sky for the benefit of mankind. Yet that simple statement seems to glorify the God of creation, *"And He made the stars also."* The stars are constantly becoming non-existent, while others are being born. Stars are actually dense clouds of gas that are being formed from the sun. When stars run low of inner hydrogen fuel, they become unstable and begin to release their outer atmospheres in spectacular and often beautiful ways **(18).** The more recent development of the Hubble telescope has allowed scientists to witness the formations and disintegration of stars.

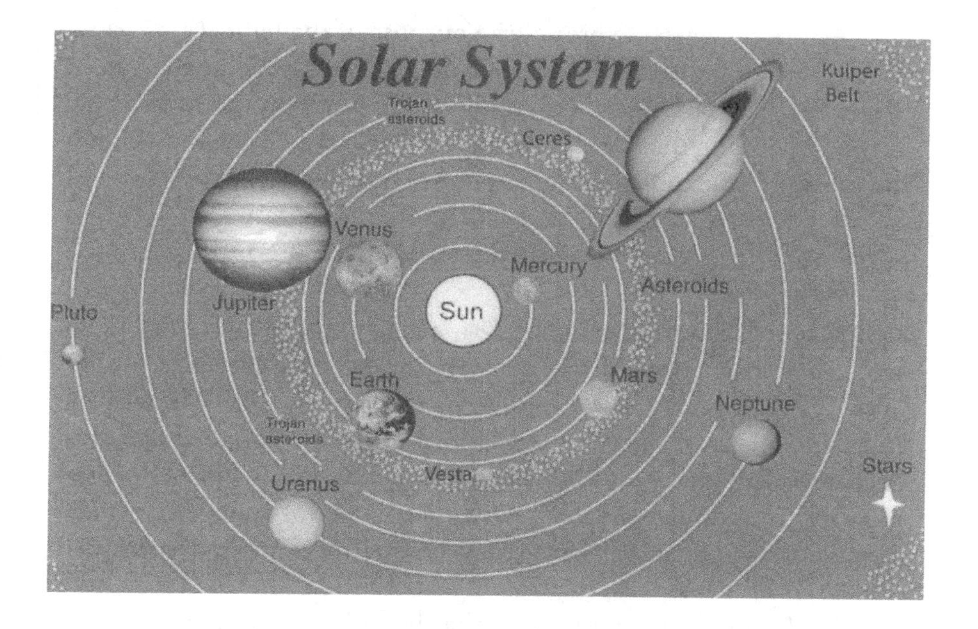

As one studies the vast scope of the size of God's creation, he has to be amazed at how small a person is in comparison. Even the oceans are so massive that man cannot comprehend their size, much less think of how much more awesome is the Creator, who created all of it. The blessed part of this is the fact that this great and mighty God is interested in every small creature, but especially man. The **Psalmist (147:4)** declares God "*counts the number of the stars; He calls them all by name.*" Sir James Jean tells us there are more stars in space than there are grains of sand on all the seashores of all the world **(19).** As one studies the size of God's creation, he should conclude that it is far too vast for man to conquer or even to discover. A person who seeks the truth should be drawn to find the Architect that designed the universes. Each planet and each star hangs on nothing, so exactly how does that happen in a chance environment? The planets never get out of orbit and never crash into each other. This alone should verify the fact that someone is keeping all things according to their design. Only the One who made such a sophisticated planetary structure could possibly control its function.

Again, the writer expresses the purpose of the lights **(v. 17–18)** to give light to the earth. The sun would rule the day, while the moon would rule the darkness. God approved His work on the fourth day as good by stating the fourth time that *"it was good."* The evening and the morning concluded the fourth day of creation **(v. 19).**

Fifth Day

Day five begins in same pattern as the previous days with God giving a voice command and speaking things into existence out of nothing. God said, *"let the waters bring forth"* **(v. 20).** The waters and the firmament obeyed the Master's voice as it supplied all sorts of animals in the seas and birds of feather to fly in the open sky. The waters were abundantly inhabited with a variety of aquatic creatures throughout the entire earth's crust. Whatever the Creator desired was created to fill the seas on day 5.

God brought forth creatures that had life within themselves to move from place to place and reproduce. The trees, the grass, and herbs had life, but not like the animals with mobility. The animals of the sea and the birds of the air are closely kin as they were created on the same day. These creatures did not both come from the same creatures as many would contend today as one set of creatures of all kinds are found in the waters, while the other species with wings fly through the heaven. Admittedly some creatures such as ducks, geese and others fly through the air, yet these may swim in the waters. They all are distinctly different, but do not possess the same ancestor creature.

This marked the second time that the verb created **(bara)** is used as God produced life from nothing. This has always presented great problems for the evolutionists as how that life got started. Since they can't accept God's version, so they have to attempt to explain how life occurred without an intelligent being. Evolutionists have no clear answers, so they often muffle their feeble response into billions of years of primordial soup that belched out a single cell organism called an ameba that could actually move from one place to another. That single cell organism stumbled upward into multiple celled animals and eventually became a man, according to those that share this concept of blind chance.

It is simply absurd to believe that all the myriad forms of life known with their complexity all evolved from some single cell organism. There is not a single shred of scientific evidence that has ever been produced to support such an incredible story. It seems that someone over these past billions and billions of year would have found at least one factual link in this mysterious chain of evolution, if it were true. In fact, there is at least one creation scientist, Dr. Kent Hovind that made a long time standing offer of $250,000 for anyone at any time that could offer him one piece of empirical evidence (scientific proof) of evolution **(20).**

In the Bible version, God did not just create a simple fish nor bird, but literally filled the seas and firmament with creatures of all sorts in one day. The writer specifically describes animals "abundantly" **(v. 21)** with many offspring from each species of animals within day 5. There are over 800,000 different insects, 30,000 different fish, 9,000 different birds, 6,000 different reptiles, 3,000 different amphibians, and over 5,000 different mammals **(21).** Nowhere is there any evidence of an animal that has ever changed to another kind. Although there is plenty of evidence that many species of animals have become extinct, there is no evidence of a completely new kind being formed, especially one being birthed from another kind. In fact, there are laws of genetics that rule and determine the appearance of each offspring in every class of animals. Inheritance always supersedes the environmental factors and conditions.

Sixth Day

Once again, the activity of day 6 resulted from the Lord's command of *"Let the earth bring forth..."* **(vs. 24).** On this day the larger animals were called into existence and from a shred of nothing they appeared and immediately populated the whole earth with their offspring. Cattle, beasts and creeping animals would include a large hosts of animals of all sizes and shapes. The higher forms of animals were definitely created upon this day. No doubt, this includes the dinosaurs and other mammoth-size creatures. Today there are many inquiring minds about the dinosaurs and exactly when they roamed the earth. The fact remains that if a person believes the creation story of the Bible, then God created them on day 6. By accepting this truth, then that means that man and dinosaurs lived side by side upon the earth without a huge fear of one another. Many scientific facts have already been discovered that document this co-existence on earth. Fossils of man and dinosaur footprints have been discovered close together in the same layers of rock in various locations around the earth.

In the story of Job, there are many accounts of these large sized animals that lived on land, as well as, large sea creatures. Job does not use the term dinosaur, since that word was not coined until 1841 by a famous paleontologist, Sir Richard Owen. The word dinosaur means "terrible lizard" and refers to a giant reptile like creature which lived on land rather than in the water **(22). Job (41:1–34)** describes a huge land living creature or dragon that breathes out fire from his mouth and smoke from his nostrils. This creature called Leviathan was enormous in size and strength, which no man could tame. Compared to Leviathan, the sea seems like only a small pot of oil.

Job (40:15–24) also describes another enormous size animal called Behemoth that lives mostly in the rivers, which could drink up a river **(40:23).** This iron like creature ate grass and possessed a huge tail that resembled a large cedar tree. Both of these creatures were mammoth sized animals that probably resembled the lizards of today. Some have fathomed the earth as some type of planet like Jurassic Park, but maybe there was no comparison in quantity nor appearance of those depicted in the movies.

Many evolutionists today claim that birds are the modern version of the dinosaurs, which some suggest evolved into birds. Artists further draw pictures depicting reptiles with feathers and avian features to support this phenomenal transformation **(23).** These same evolutionists teach that these animals lived some 235 million years ago and long before the coming of man **(24).** If one believes the Bible account of creation, then there was nothing physical prior to God speaking the planets into existence during the week of creation.

According to the Bible record, there would have been nothing physical until 6,000 to 10,000 years ago. Thus, this account of these dinosaur-like creatures could not have existed 235 million years ago, if the planets did not exist. A person must decide which story is true, but there is no way that that evolution can be reconciled with the biblical account as some contend.

It seems unlikely that few, if any, of these great size animals are still living on earth today. Most likely, these were important to God's original creation for the checks and balances of nature as they consumed large quantities of green food. Considering that the subtropical climate and temperatures were just right to produce large size plants and trees. There was a greenhouse like setting where an oxygen rich environment was placed under the water vapor canopy sealing out all harmful sun rays. Plants and animal life reached their full potential in this perfect setting until the Great Flood of Noah's day came destroying the protective water vapor canopy in the upper atmosphere. With the breaking of the water vapor canopy, the oxygen content of the atmosphere surrounding the earth drastically changed for the worse.

After the creation of the higher forms of animal life, the Lord came to the crowning jewel of all of His work in the creation of mankind. All things were set perfectly into their place before He brought man upon the scene. The Lord rolled out the red carpet for the choice and highest order of His act of creation. He literally filled the earth, the seas and the air with animals of all kinds after He had clothed the earth with a full complement of plants prior to the creation of man. All the necessities for living creatures were present on earth by this time: light, air, water, soil, chemicals, plants, fruits, and so forth **(25).** The Creator had an ingenious plan in place before He began to speak all things into existence, as He did nothing as an afterthought. Since God is an omniscience Being, then absolutely nothing was left to blind chance.

Nowhere does evolution come into conflict with the Bible more than at the point of God's creation of the first man. The term "evolution" is a general theory that all life on earth has evolved from nonliving matter and progressed to more complex forms with time **(26).**

This particular view has had a massive influence on the sciences, humanities, theology, and government with increasing evolutionary assumptions **(27).**

God's record of creation states that man was created in the image of God, but exactly, what does that mean? Since God is not physical and has no human form, it obviously is a reference to the spiritual nature of man that resembles the Spirit nature of the Supreme Creator. John declares that *"God is Spirit"* **(4:24),** and in God's creative act in Genesis, he created man in His image, which refers to his spirit life. The second time the verb "bara" (vs. 21) is used to create life, whereas, in 1:27, it refers to the creation of spiritual life. Man is the only creature that is kin to God because he possesses a spiritual life like unto his Creator.

The atheists and others would have man to believe that he ascended from animals and thus he is akin to animals. However, man possesses intellect, a conscience, a will, driving emotions and creativity like unto his Creator. Not one single animal possesses any of the aforementioned characteristics that are found in the least educated man that exists on earth. There is not one animal that can change its environment, yet man can develop ways to heat and cool his surroundings. Man can cook and prepare his foodstuff, but no animal ever cooks its food. What animal has the gift of speech? Even the most primitive human tribe possesses linguistics of a subtle, complex, and eloquent nature **(28).** Man is not related to the animal kingdom and he surely did not descend from them as the secular science world would have people to believe.

True science always deals with facts that have been observed through study and laboratory testing is frequently involved. However, in the case of evolution, no person has ever observed the change from one species to another. There are no facts, no observations, nor experiments to make this theory a form of science. Thus, evolution is not really a science at all; it is a philosophy or an attitude of mind **(29).** Evolution must be believed that makes it a subject of faith, which in reality is a religion and not a science.

Not only is evolutionary philosophy basic in most anti-Christian social, economic, and religious philosophies, but it is also the pseudo-scientific rationale of the host of antisocial immoral practices that are devastating the world today such as: abortion, the drug culture, homosexual activism, animalistic amorality, and so on **(30).**

Recently, someone shared this interesting poem entitled "The Monkey's Disgrace"

> Three monkeys sat in a coconut tree,
> discussing things as they are said to be.
> Said one to the others, "Now listen you two,
> there's a rumor around that can't be true.
> They say that a man descended from our noble race.
> Why the very idea is a great disgrace!
> No monkey has ever deserted his wife,
> starved her babies and ruined her life!
> And you've never known a mother monk
> to leave her babies with others to bunk!
> Or pass them one to another,
> until they scarcely knew who is their mother!
> Here's another thing a monkey won't do,
> and that's go out at night and get a brew!
> Or use a gun, a club, or knife
> to take some other monkey's life.
> Yes, man descended, the ornery cuss,
> but Brother, he didn't descend from us.
>
> **(31)**

God gave man an assignment of rulership and dominion over the animal kingdom, which places man in a superior position over all those animals that God created on day number 5 and 6.

The statement *"Let us make man in our image"* **(vs. 26)** is often puzzling to many students of the Word of God. Why are the pronouns plural instead of singular. Some have suggested that this is an early reference to the triune Godhead of the Father, the Son and the Holy Spirit. However, the Hebrew writer Moses likely had no knowledge of such persons of the Godhead and certainly there was not a heavenly counsel of the three convening to decide on how to make man.

Since the Bible clearly teaches that there is only one God, this explanation is not acceptable. The earliest records of writings show that the Hebrews would very often shift the pronoun references of their Deity from singular to plural, depending upon the emphasis. For example, when they wanted to show the majesty of their God, they would make the pronoun plural, while on other references it would be singular. Only the New Testament writings present the picture of the Triune Godhead, thus, the plural pronouns present the case for a majestic monotheistic understanding of the Creator God that the Hebrew writer held. The word "God" in Hebrew is transliterated as "Elohim," which is a plural word, yet He was a single God. Even though the people of old may not have had the concept of a triune Godhead, the concept was already set in place in Genesis 1:26.

The Lord multiplied each kind of animal that He made on day 5 and 6, but He only made one male and one female of mankind and commanded them that they were to populate the earth. This is a clear indication that all mankind have come from this one man Adam, including his own wife Eve, who was taken from Adam's side. This teaching is very important to understand how the redemptive plan of the Lord works as the sin nature of man is passed on biologically from one generation to another. Once sin entered into the world, there is no reversing the course and putting this evil genie back into her bottle. Since sin carries a payment of death, there was no person who could satisfy that payment. Only the Second Adam could redeem mankind and pay the high price for sin and satisfy the price that God required to rid sin from the life of a human being.

God said to Adam and Eve, *"Behold, I have given you every herb bearing seed, which is upon the face of all the earth, and every tree, in the which is the fruit of the tree yielding seed; to you it shall be for food or meat"* **(1:29).** God supplied food for man from the fruits, vegetables and the vegetation upon the earth and not meat of the animal kingdom. In fact, God forbid even the animals to eat the meat of another animal **(v. 30)** in God's perfect environment. The animals were given the herbs and the grass to eat for food. Remember that this was God's perfect plan in the garden for man and the animals to be vegetarians.

The animals would live in perfect harmony with each other where the lion would lie down with the lamb. Also, man would live peacefully with the animals and there would be no fear between the animals and man. With the greenhouse environment, the earth produced plants, herbs, grasses, fruits and vegetables for all to thrive with plenty of food for all of God's creatures. It was not until the water vapor canopy was broken during the flood that the food of plenty disappeared. Following the Flood **(Gen. 9:3),** God permitted man to eat the animals and for animals to eat other animals. Man was forbidden to drink the blood, which contained the life which was to be poured out on the ground. This indicated that God's perfect bond of peace, love, trust had been broken by sin.

After God completed His work of creation in the sixth day, He marveled at all that was made and created **(v. 31)** like some great artist standing to gaze at his painting. Then the Lord said that His work, *"was very good"* **(v. 31)** The Lord said that it *"was finished"* **(2:1),** which is the very same thing that Jesus said while hanging upon the cross, which meant that this phase of His creation was finalized.

Seventh Day

Then on the next day following the sixth day, God rested from that first phase of work that He had been doing. God was not weary nor tired, since He never becomes fatigued or worn out from work. Obviously the God who can create more universes than man can count, and who can lock up within the tiny breast of the atom enough energy to obliterate an island, cannot possibly grow tired **(32).** This was a valuable lesson and example that the Lord set for all mankind. No man is designed to work continuously without cessation from his labors. God knew that man would need such a day, but God does not. Later the Lord would establish this extra day for His people of the faith as a Law to be kept by all His people. He *"blessed the seventh day"* **(2:3)** even from the beginning as that special day for all mankind and set it apart from all the other days as He *"sanctified it"* **(2:3),** which means "cleansed" and "set apart" from the days of work.

There is a wonderful insight given by the writer of the book of Hebrews about the "rest" of the Lord regarding an entry into the rest of the Lord. The writer declares a promise has been given for the rest of the Lord in case any follower might have any misgivings over receiving this special time with the Lord **(Heb. 4:1).** This period of rest is definitely connected to the seventh day of rest as the Hebrew writer says, *"Although the works were finished from the foundation of the world. For He has spoken in a certain place of the seventh day in this way: And God rested on the seventh day from all of His works"* **(Heb. 4:3–4).** It is abundantly clear that only true Believers will be able to join God in this period of rest. However, the Lord welcomes all mankind to join with Him in a special time every week, which will give all followers a little foretaste of the eternal rest of heaven. The six days of creation completed or finished a perfect creation that would last forever and ever. The Creator knew that He would have to destroy that perfect creation because of the sinfulness of mankind, which would lead to a fallen world that would be far from perfect. This ended the first creation story.

Basic Elements in the Creation Account
(Ross, Allen P. **33)**

DAY	**ACTIVITY**	**THEME**
1	Light created; Light divided from darkness	Goodness separated from evil.
2	Heavens divided from waters below An expanse called a Firmament	Separation of waters
3	Surface of earth covered with vegetation	Fertile land for vegetation and plant life covering all the earth.
4	Lights placed in sky for time and seasons	Order of dating, time and seasons.

5	Living creatures placed in sea and sky	Multiple Sea creatures and fowl
6	Land animals of all sizes and kinds Human being was created in image of God	All kinds of land animals Pro-creation and dominion for all.
7	Designation for a day of rest	Provision for rest and sanctification

Chapter 7

The First Creation Story

1 Morris II, ***The Genesis Record***, pg. 37
2 ***Ibid***, pg. 38.
3 Barnes, Albert, ***Barnes Notes of the Old and New Testaments,*** pg. 40.
4 ***Ibid.***, pg. 33.
5 Marsh, Spencer & Seelig, Heinz, ***Beginnings-Portrayal of the Creation***, Multnomah Press, Peabody, Maryland: 1981, pg. 19.
6 ***Ibid.***
7 Morris II, ***The Genesis Record***, pg. 59.
8 Ham, Ken, Snelling, Andrew, & Wieland, Carl, ***The Answers Book***, Answers in Genesis, Master Books, El Cajon, Calif: 1992, pg. 120.
9 ***SBC LIFE***, Southern Baptist Convention quarterly magazine, Nashville, Tenn., quote from ***US News and World Report***, January 1998 issue, pg. 12.
10 Thieme, ***Creation, Chaos and the Restoration***, pg. 10.
11 Bradstreet, David & Rabey, Steve, ***Star Struck,*** Seeing the Creator in the Wonders of Our Comos, Zondervan Press, Grand Rapids, Michigan, 2016, pg. 208.
12 ***Ibid.***
13 ***Ibid.***
14 ***Ibid.***
15 ***Acts & Facts Magazine***, Institute of Creation Science, October 2018, pg. 10.
16 Bradstreet, ***Star Struck***, pg. 207
17 MacArthur, John, ***The Battle For The Beginning***, W Publishing Group, Division of Thomas Nelson, Nashville, Tennessee: 2001, pg. 106.
18 Bradstreet, ***Star Struck***, pg. 208.
19 Phillips, John, ***Exploring Genesis***, pg. 44–45.
20 Hovind, Kent, ***Creation Science Evangelism***, A resource supplement Creation Science, Pensacola: Florida: 2000, pg. 7.

21 Phillips, ***Exploring Genesis***, pg. 44.
22 Ham, ***The Answers Book***, pg. 21.
23 Gallings, A. P., ***Dinosaurs,*** A Pocket Guide, Answers in Genesis, Hebron, Kentucky: 2010, pg.
24 Horner, J. and Lessem, D., ***The Complete T Rex***, Simon and Schuster, New York, New York: 1993, pg. 18.
25 Morris II, ***The Genesis Record,*** pg. 68.
26 Ankerburg, John, pg. 5.
27 ***Ibid.***, pg. 6.
28 Phillips, ***Exploring Genesis***, pg. 45.
29 Morris, Henry III, ***Science and the Bible***, Moody Press, Chicago, Illinois: 1951, pg. 35
30 ***Ibid.***, pg. 41.
31 Hovind, Kent**, *Creation Science Evangelism***, pg. 37
32 ***Exploring Genesis***, pg. 41. pg. 47.
33 Ross, Allen P., ***Creation & Blessings***, Baker Books, pg. 75.

CHAPTER 8

THE SECOND CREATION STORY

Gen. 2:4–25

The second account of creation begins with a statement of the era when *"the heavens and the earth were created"* **(2:4).** The often asked question is why a second creation story? Is this a clear indication that there were multiple authors rather than just one writer? Many suggestions have been offered concerning the multipliable author view, but none of them seem to be quite satisfactory, since the main argument is over the varying terms for Deity of "God" used in the first account and "Lord God" used in the second creation account. See the discussion earlier on authorship in chapter 2. The writer is given more details in creation in the second account, suggesting that some period of time could have elapsed between the writings. Moses uses the term "generation" which means family history; however, he is referring to the history of the heavens and the earth.

A new name for the Divine Creator is introduced in the very first verse of the second creation story as the "Lord God." The term "Lord" is "Jehovah," which is the redemptive name of God and is connected to the name "Elohim" to form a covenant name that will be used eleven times in the second story. In the first account, God was the Creator of the universe, but now He is seen as the Creator and Redeemer of all parts of creation.

In the first account, God created the *"heavens and earth,"* but now the focus becomes the earth and not the universe as the phrase is reversed to *"the earth and the heavens"* **(2:4).** Among all the universes and planets that God created, earth would become the primary place that He chose to reveal Himself to man. The story of God's great love is still being constantly revealed, like some unfolding drama. The Lord God created all the planets that exist, since Paul declared to the Colossian church, *"All things were created through Him and for Him"* **(Col. 1:16).**

The writer reveals that before the worlds existed and far beyond the plants of the fields that the Lord God was there taking care of the business of the universes before it had ever rained **(2:5).** This account clearly agrees with the earlier version that the plants and animals came before the creation of man.

There was not any rain upon the earth, but God had a far better method of a fine mist that went up from the earth. This method eliminated soil erosion, ravaging rivers, flooding problems and drainage problems. This was the ideal and perfect environment that God had planned, so no harmful matters existed. It is obvious that the writer of the story had already witnessed rain upon the earth, which did not occur until the flood. The mist that rose up from the ground performed as a perfect irrigation system for all the earth and its vegetation.

"The Lord formed man" **(v. 7)** from the earth, thus he was a product of the previous materials already made by God. However, there was a missing ingredient that required the breath of the living Almighty God. It has often been illustrated that the same chemical composition and minerals found in the human body exist in the topsoil. Even though, there is the availability of all the ingredients for human life, no one has been able to produce a human by utilizing all the resources. The problem is that life requires a breath of the Divine God who is the Creator of life. Life is far more complex than the physical elements found within a living being. Adam rested upon the newly created ground on which he lies that he himself, in his still dreaming existence, is strange and marvelous to the highest degree, but just the same he is a piece of earth **(1).**

In addition to the need for God's touch upon any substance or matter to produce a living cell, human life is composed of a spiritual dimension that the animal kingdom simply does not possess. This verse **(v. 7)** clearly teaches the concept of a spiritual being that is like unto the Creator, Himself. Man does not possess a soul, but *"man became a living soul."* By the power and design of God, man became more than an animal by the special touch of a Master Designer. This singular statement cancels out the theory of evolution and destroys the teachings of life beginning with a single cell that moved upward forming all the species of animals. Creation and evolution are totally at the opposite ends of the poles of truth and error, yet the evolutionist insists that man was not designed as a being but came into existence by a series of chances. The evolutionist always preys upon man's ignorance to truth and his gullibility to fascinating fictional stories.

The Lord God brought that first man into a beautiful, bountiful and fruitful place called "Eden" **(v. 8),** which means paradise. There was no more complete place in history than that place where God established for the first family. It was unsurpassed in beauty and splendor with an endless supply of foodstuffs. No one would ever have a want that could not be met in the perfect garden called Eden.

Not only did the garden provide food that was good, but it was a beautiful place with pleasant **(v. 9)** sights and sounds. In the midst of the garden **(v. 9)** of plenty stood two trees that were as different as daylight and dark. These trees signify the two opposite destinies that await man as he would have the option of choosing one over the other. The first tree was the "Tree of Life" or "Eternal Life" that would be offered to every human being who came to the garden, while the second tree offered death and suffering. The Tree of Life was the first tree to become extinct on earth. The significance of the Tree of Life, of which so remarkably little has been said earlier, is only really comprehensible here. Indeed, it is now obvious that the whole story has really been about this tree **(2).** That Eternal Tree of Life must not be ignored, since it is at the heart of the Gospel message. It is interesting that it would take Jesus being nailed to another tree to accomplish the eternal plan of redemption for the sin that began by eating the fruit from that wrong tree.

Even though the Eternal Life Tree which represents Christ was temporarily hidden to human sight, it reappears in the New Testament when Jesus declared, "I Am the vine and you are my branches" **(John 15:5)**. Jesus declares Himself as the source of Life and if man is to possess life that is Eternal Life, then he must be grafted into that source of life. Jesus said, *"without me you can do nothing"* **(John 15:5).** Apart from Christ, man like a pruned vine or a tree limb will quickly wither away.

Man was thus offered the way of life through God's offering of a tree in the garden or man would be free to choose a deadly alternative with another tree that looked beautiful, but was tragically filled with deadly poison. It would provide what man could obtain upon his own, which was a bad, bad choice. The Creator, in His infinite wisdom had to offer a plan for man to choose, if man were to be more than a puppet on a string. This alternate tree was called the tree of knowledge of good and evil. Adam was permitted to eat of everything in the beautiful garden, except from this one tree. God's command stated that Adam was not to eat of this tree when He declared *"for in the day that you eat of it you shall surely die."* (Gen 2:17) The choice for man would result in a moral choice between what God called good and that which God deemed deadly and dangerous for man. Man could choose God's way which would lead to an eternity of love and blessing, or else he could choose to get what he could obtain upon his own with his own wisdom and knowledge.

God not only created a beautiful garden, but He sent forth a river of water from the midst of the garden to supply the world beyond with life. It was a most unusual river that divided itself into four parts to supply each corner of the earth. At the time of the writing, people believed that the earth was flat with four corners. The river seems strangely akin to the pure river in **Revelation (22:1)** that flowed from the throne of God and the Lamb. In that setting, the Tree of Life is again described standing beside the pure river with twelve manners of fruits, which it bears in every season.

The river from the throne of God proved to be essential for all of life, just like the Lord God is essential for all of life. Jesus Christ, the "Living Water" that supplies every need of man is the substance of the River of Life that flows to every part of the world.

The river of life was the source for the Pison River (Indus River), which flowed toward India. The second river was called Gihon in the land of Ethiopia, which is traditionally known as the Nile River. The third river is called Hiddekel and flows toward the east of Assyria, which is all likelihood is the Tigris River. The fourth river is the Euphrates River. These four rivers were the most famous rivers from the ancient world, so the writer chose to use these rivers, which supplied water for a major portion of the known civilization in those days.

The Lord God placed Adam in the Garden of Eden **(2:15)** to be the gardener and guardian for God's creation. If man is to be fulfilled, then he cannot remain idle. Thus, the Lord provided him a challenging occupation that would be a pleasant and an enjoyable task compared to his predicament after his fall into sin. Adam was faced with a meaningful life of caring for the garden, but also given one single prohibition as he spent his life in the paradise setting. He was forbidden to eat of the fruit of the tree of knowledge of good and evil.

The Lord commanded Adam **(2:16)** to be trustee of God's creation and permitted him to eat freely of all the trees of the garden except that one tree of knowledge. This happened to be the only negative commandment that the Lord charged Adam to keep. Everything else the Lord made for man's enjoyment and pleasure, thus withholding nothing from him that was good. This statement perfectly parallels an earlier statement the Lord made in the first creation account **(1:28),** when He said *"have dominion over"* and *"subdue it."*

Man was asked to trust God and not eat of the fruit of only one tree or else it would spell his doom and death **(2:17).** The penalty or the punishment for disobedience would be serious as man would experience death. This was not a threat; it was a simple statement of fact that is still true today, as the punishment for sin is death **(Rom. 6:23).**

There was clearly a choice given to Adam. He could enjoy the bounty and the blessings of life in paradise with plenty of food and prosperity or he could have a death sentence cast upon his life by disregarding God's word. Man had a choice to believe God and obey Him or to do what God had forbidden regarding the Tree of Knowledge of good and evil.

There was one and only one thing in all of creation that the Lord made that was not good. That was the fact that Adam was alone with no compatible mate for life. None of the animals could afford Adam any meaningful relationship, nor was there found anyone that could relate to Adam in human terms. The Lord God desired that all of man's needs would be met physically, emotionally, spiritually, so He designed a wonderful partner to share life with man.

This second story differs somewhat as the Lord formed man; He designed and manufactured each animal in creation. Even though, the telling of the story in the second creation account follows the creation of man, there is no contradiction in the word of God. The sequence of creation events is already carefully detailed and established in the first creation story, so the writer is more interested in relating the account of the mate for Adam.

For the first time, the author calls man by the proper name of Adam, which means "the man." God took the dust and formed and fashioned each of the animals that makes up the vast domain of the animal kingdom. Then God presented the animal kingdom to Adam and allowed him to give each one a name **(2:19–20),** but not any supplied the deep need of Adam's life for human compatibility.

Then God caused (**2:21**) a deep sleep to come upon Adam, which was no ordinary or natural matter. This sleep was much of the order of an anesthesiologist that would alleviate any pains or discomfort, while the Lord God removed a section from the side of Adam including a rib bone. After the Lord performed the first surgery on man, he immediately healed up His patient because he is the Great Physician. He then took the rib and made a complete woman from that one bone. God can always take a little and make much from it. In fact, He can take nothing and make something as He did in **Genesis 1:1**. However, it is critical for redemption of the human race that all mankind come from one source in order that all be linked together. Thus, all mankind came from this one man Adam. The biblical account clearly declares that each person possess a common ancestor that is in no manner related to a monkey and leaves nothing to chance as God continues to oversee each act of creation of man and beast.

The creation of the woman was a highly irregular happening, since the woman received her being from the man, rather than the normal occurrence of a woman giving birth to the man. Remember that God always does the supernatural that cannot be accredited to the work of man. This event literally baffles the atheistic version for the origin of life, since it cannot fit into their scheme of events.

Perhaps, by the time that Adam awoke from his deep surgical sleep, God presented him with the best gift that he could have ever dreamed of in the person of a woman. Adam's delight was immediate as he looked upon God's perfect work of creation in his helpmeet. He certainly was made aware that his mate had been formed from his own body and broke forth in a song of great joy. He rejoiced that she was bone of his bones and flesh of his flesh **(2:23).**

Some have suggested that the story of God creating the woman from Adam's rib is merely mythological. If it was indeed a myth, then Jesus was deceived since our Lord related it as historical and factual **(Matt. 19:3–6).** Once again, the Bible is filled with truth on every word and not one single word is out of place, as the Bible is a perfect book with all things in order.

The writer stated that God took the woman from man's side. Matthew Henry says that the woman was taken from Adam's side, not from his head to rule over him, not from his feet to be trampled on, but from his side to be equal with him, from under his arm to be protected, from close to his heart to be loved **(3).** The rib bone is the only bone of the human skeleton that could be removed without obviously impairing the mobility and appearance of Adam.

The Creator had put together a perfect marriage union in a perfect garden setting of paradise and pronounced wedding vows **(2:24)** for a oneness in that relationship. God planned all the details, as well as, the partners, so it was without flaw. He always does perfect work and that first union embodied the holiest of ideals for all future generations. The concluding statement declares that they were both naked and not ashamed **(2:25)** before God nor each other, which clearly emphasizes their innocence and moral purity. This is a virtuous statement as holy marriage before God is a great virtue. God has always demanded and expected His people to remain pure when they arrive at His altar for marriage. Nothing less than moral purity is acceptable before the Lord God. Two people, male and female have the greatest privilege of becoming one in the sight of the Lord, while living in peace and harmony before their Creator God.

Chapter 8

The Second Creation Story

1 Bonhoeffer, Dietrech, ***Creation and Fall***, A Theological Interpretation of Genesis, McMillian Publishing Company, Inc., New York, NY: 1978, pg. 47.
2 Ibid., pg. 89.
3 Phillips, ***Exploring Genesis***, pg. 53.

CHAPTER 9

THE CURSE UPON THE UNIVERSE

Genesis 3:1–24

The teachings of evolution will be dealt a huge blow in this chapter as man is described in the Bible on a downhill plunge, rather than improving over time. Those who adhere to the principles of blind chance evolution, believe that the animal kingdom is continuing to climb upward in a higher and higher position. However, the Bible contends that man began in a perfect setting with an ideal beginning, yet he began a plunge to a much lower level in civilization throughout almost each succeeding generation. Sin brought a curse upon man, the earth and all of God's creation that cannot be reversed by any man. Mankind continues to be plagued by the sin of his life as it was for that very first family in the garden. This pattern of sin shall continue to be passed on in what could be called spiritual genetics of mankind.

The scene from the previous chapter of the second creation account turns dreadfully into a downhill disaster for God's first family of paradise. The first two chapters of the Bible gives no hint of wrong, sin or the devil; however, all that immediately changes in the opening verse of chapter 3. The tranquility and beauty of paradise is forever changed by a powerful, perverse, and persuasive character who is far wiser *"more subtle"* **(3:1)** than any creature that God made upon earth. Apart from God's help, mankind is up against a clever devil that can outwit or deceive the smartest human being on earth. Without God's help, man will be a loser every time he engages in spiritual warfare with Satan.

The new creature that suddenly appeared in the Garden of Eden was a fallen angel who rebelled against God in Heaven in order to set up his own kingdom that would rival and then replace the Lord God. Since the limited power of Satan was no match for the powerful God of Creation, this perverted leader was cast out to planet earth. He will never be able to walk into Heaven again and disrupt the Kingdom of Heaven. So sin did not begin on earth, but rather in Heaven when Satan rebelled against God **(1).** Satan came walking in the Garden of Eden after he was cast out of heaven to afflict God's creatures and cause them to rebel against the authority of God's Word. From the picture given by the writer, Satan came strutting into God's earthly paradise to destroy the creatures that the Lord God created that were in the image of their Creator. Since the devil had not succeeded in heaven, then he would destroy God's prized creatures on earth. Ezekiel **(28:13–17)** declares that he was a creature of magnificent beauty and wisdom that had been an anointed cherub in heaven.

Since there were no sinners on earth to work through, Satan had to work through an animal to accomplish the fall of man, who was made in the image of God. So the devil carefully chose one of the willing and maybe the best looking of the Lord's creatures. The snake surrendered to be a part of that plan to bring about the fall of man and it was truly a debating fall for all humanity that followed. In contradiction to much popular theology, man is not on the way up. He has been on a downward spiral of destruction since the scene in Genesis 3 with a basic nature to pursue the way of evil that would ultimately lead to their disbandment from the perfect Garden of Eden in which man was created to inhabit for an eternity on earth.

The serpent embodied the spirit of the fallen spiritual leader that had become an enemy to God and to all His creatures of creation. (Please refer to chapter 5, which deals with a study of Satan.) His first statement to the woman was a question that was filled with an insinuation that God was not good and that He was withholding good things from her. It was a very untruthful statement from the master of deception that would cast a cloud of doubt in the mind of Eve.

The woman answered the adversary with the Word of God **(3:2–3)**, which should be every Believer's reply. She was not ignorant of God's Word and knew exactly what God had declared. Eve explained to the devil that only one tree was off limits, which would carry the death sentence, if she ate it. At that point, she completely trusted God and His Holy Word. However, she made a serious mistake by continuing the dialogue with the devil rather than turning away. The first woman did not know to resist this creature of darkness and deception. James declared in the Word of God *"Resist the devil and he will flee from you"* **(4:7).** Perhaps Eve believed that he was simply misinformed about God's Word and she would be the one to help him understand the truth about God's Law. Satan had literally challenged the authority of God's holy Word, as well as, its accuracy and application.

The continuation of the conversation between Satan and the first woman lead to the devil making an outright declaration of falsehood **(3:5).** Since Jesus declared him to be a liar and the *"father of all lies"* **(John 8:44)**, he truly revealed that he cannot tell the truth, but only falsehoods. The devil had now trampled the Word of God into the dust of serious error by stating that God was a liar and He would not destroy the woman; but he added how much more complete that she would become by following his path of rebellion. Cursed is anyone that adds to or subtracts from the Word of God. Paul declared it clearly in his writing to the church at Corinth **(2 Cor. 11:3)**, when he stated, *"I fear, lest somehow, as the serpent deceived Eve by his craftiness, so your minds may be corrupted from the simplicity that is in Christ."* Paul's concern for the young immature Believers could easily be persuaded to believe a different gospel than what he had previously proclaimed to them. Beware of those that may come claiming they have found a better way to salvation.

Satan had just offered the woman the fruit of the tree that God had forbidden her to eat. However, Satan offered a vain promise that she would not die, but be equal to God, if she received it. He insinuated that Eve would have as much wisdom as God and would become a goddess. All of it sounded too enticing, so she took another look at the tree through new eyes of understanding. Eve now had just committed the first sin of doubt as she disbelieved God's Law and begin to lust for the fruit of the forbidden tree. Satan had scored a major victory in her mind as he persuaded her to turn her heart and commit the second sin of lust **(3:6)** as she *"saw."* The Bible says that she *"saw that the tree was pleasant to the eyes"* and that she *"desired"* (**vs. 6)** the fruit from the tree of knowledge of good and evil. Eve had gone from the sin of doubt to a desire to have that which was forbidden by her God.

The woman quickly ate the fruit after the desire and passion for it had entered her heart. What is so amazing is that Satan never tempted Adam as he merely ate at the invitation of his wife. He took the fruit and ate with his eyes wide open without any debate or question. It is true that sin seldom flies solo, as it seeks to bring others along the same path. Immediately after they ate the forbidden fruit, *"their eyes were opened, and they knew that they were naked"* **(vs. 7)**. Temptation often comes through the human eyes and now they saw things completely different from before. Someone has said that there is no harm in looking; however, sometimes it only takes one look for a person to become hooked on an object that can create evil in his life. There is a great truth in a children's song which declares "be careful little eyes what you see."

Satan had carefully sought out a plan of attack that would invert God's original plan for the family. Since Adam was created to be the head of the family, God's plan was to give man the intellectual leadership role in the family. However, Satan devised an attack upon the second creature—the woman to whom the Lord gave emotional leadership gifts for the family. Thus, Satan crafty plan was to appeal to the heart and not to the head. The Bible says in **Jer. 17:9,** *"The heart is deceitful above all things and is desperately wicked."* Satan successfully twisted and inverted God's order of leadership approaching the woman rather than her husband.

It should be clearly stated at this point that God never tempts anyone to sin (**James 1:13**), thus he will never coerce one to commit a sin. Also, the Lord will not allow Satan the power to force or coerce one to sin. Popular TV star of the 1970s, Flip Wilson declared that the devil made him do bad things, but that is an impossibility. The devil certainly has tremendous persuasive powers, but he cannot force his will upon anyone. Therefore, sin is a bad choice whereby man chooses, when he knows the law of God. **James (4:17)** states God's case for sin, *"therefore to him who knows to do good, and he does it not, to him it is sin."* Eve knew God's law and she knowingly made a bad choice. Once this thing of sin had been committed, things would never go back to the way they were intended. The evil genie was out of the bottle. Just like squeezed toothpaste will not go back into the tube, things for all of mankind would never be the same.

When God created man in His own image, the Lord desired to give man the freedom to choose. Man can make the choice to sin and he can make the choice for God's offer of salvation, since both are a choice that man can make. The idea that man was made in the image of God gave man a privilege of choice for right or wrong. No other creature, except man has this privilege within their nature.

The term "naked" would indicate a sense of guilt, which usually accompanies the commission of sin. It leaves the victim feeling incomplete, unclean, ashamed, and attempting to conceal his condition. Adam and Eve found themselves making a feeble attempt of covering up the unclean part of their lives. Since sin is an act of the flesh, so the flesh needs to feel better about its unclean state. The first couple found fig leaves that were large-sized leaves available in the garden which they felt would cover their problem of sin. Before the fall, Adam and Eve had been clothed in the garments of light, but their sin had extinguished the light of God by separating them from that source of God's light. Adam and Eve allowed the physical aspect of their lives to replace their spiritual sensitivity with which they were created. So they sought a physical remedy for their inferior feelings of shame and guilt. Obviously, they had not yet learned that man's remedy and coverings for his sin problem are always insufficient.

"They heard the voice of the Lord God" **(vs. 8)** assures man that God takes the initiative to go searching for those caught in the web of sin. Jesus told the story of the Good Shepherd that had a hundred sheep and when he found one was missing, he went to search for the one that had gone astray (**Lk. 16:4**). The Lord went searching for Adam and his wife in the beautiful garden, likely where they had previously enjoyed each other's companionship. God called his creatures to Himself, just as He continues to call sinful men unto Himself. The sound of God's voice was an unwelcomed sound to Adam and Eve upon this occasion. It was not the same as before, so they chose to hide themselves in an out of the way place in hopes that God would pass them by. Perhaps, at that time Adam and Eve had not come to realize that God sees all things all the time. When God called out, *"Where are you?"* **(Gen. 3:9)** it must have been like an arrow shot through the hearts of Adam and Eve. They were not ready to openly deal with their sin and face the consequences of disobeying the Creator.

Adam knew that he had been found out **(vs. 10)**, so he began to explain to the Lord about his naked predicament. Then God confronted Adam with a probing second question of *"who told you that you were naked?"* Quickly on the heels of the second question, the Lord asks a third and even more pointed question, *"Have you eaten of the tree?"* **(vs. 11)** Adam should have said, *"I have sinned before Heaven and in your sight and am no more worthy to be called your son"* **(Lk. 15:21).** Instead, Adam began to make excuses and alibis that had caused him to eat of the forbidden fruit. Almost everyone has used this same approach at one time or another to excuse themselves for a bad choice of actions or giving in to a temptation to sin. Adam actually blamed Eve for his fate and then Eve quickly passed the blame to the devil, who had caused the whole problem.

The Punishment for Sin

God turns from the man and the woman to deal with the devil in verse 14. The Lord asked no questions of the devil nor offered him any moment of defense. Because of his participation in the scheme to tempt man, the devil would be sentenced to the lowest form of animal upon the earth. The suggestion is that the serpent was a beautiful creature that walked tall upon the earth; however, his curse would force him to crawl upon his belly in the dust of all other creatures in a despised role.

Adam and Eve listened as the Lord God gave the first statement regarding a promise of their ultimate deliverance over their adversary **(v. 15)**. A prophetic promise of the gospel message was given as the Lord God declared that one would come from the seed of the woman that would crush the head of the serpent. This statement must be seen in light of the New Testament as the victory of Jesus over death, hell and the grave. Satan was given the mortal blow at the cross of Calvary as Jesus declared His earthly ministry completed by saying, *"It is finished"* **(Jn. 19:30).** That statement from the cross meant that Jesus had perfected His work of salvation for the souls of all mankind. The word "finished" can be and is translated as "perfected." All of the devil's efforts to detour, derail, and detain Jesus from His appointed mission on earth had utterly failed and now it was too late for him to have any chance to thwart God's plan of redemption. Even the devil's last flickering hope of the grave over Jesus was dashed on the first Easter morning as Christ came forth from the dead.

This comment in verse 15, is often called the first statement of evangelism found in the Bible. Even though, man had suffered a tremendous fall, there was a promised victory for a route to overcome. The Lord proclaimed a struggle and war would ensue between the devil and all mankind. However, He allowed man to see who would be the ultimate victor in the final battle.

Next, the Lord turned to the woman **(v. 16)** to pronounce her curse for having participated in sin. There would be a threefold sentence upon Eve's life. First, she could expect pain and sorrow in bearing children, which continues to extend to every woman giving birth. Second, there would be sorrow in child rearing, which would have been mere joy for the woman. Her children would rebel against her as she had rebelled against her heavenly Father. Third, the woman would become subservient to her husband as the master of her life and she would look unto man as head of their relationship.

It is very important to note that the woman was the first person on earth to receive the sin nature, thus she would pass along this nature through her offspring. Jesus would be born of a woman with that sin nature, but He would remain sinless. The Holy Spirit overshadowed Mary, making it possible that the offspring would have power to resist sin, even though His mother would have that sin nature. It is important to note that some faiths, such as the Roman Catholic believe in the immaculate conception of Jesus from a perfect, sinless mother. This is not taught in the Bible, but rather the Father of this Child would not have the sin nature and would be able to resist the temptations to sin. Thus, Jesus possessed a sinless nature from the Father, which caused Him to be repulsed by any form of sin. The Bible declares that He who knew no sin, became sin for every human being that had or will be born on earth.

God set forth four serious sentences **(v. 17)** for the future of the man. The ground that was to provide a living and substance would bear a curse that would affect man's life. Second, the thorns and thistles would bring unwelcomed problems for Adam. Weed control in croplands cost farmers billions of dollars each year in chemicals and weed control programs, in order to produce crops. The curse of that first sin has reoccurring after effects for all mankind until the end of this earth. Third, the Lord God declared that man would have to toil by the sweat of his brow **(v. 19)** as opposed to the pleasant work that Adam had been assigned in the Garden of Eden. Man would be forced to do many unpleasant tasks in order to produce his food. Fourth, the most serious consequence of Adam's sin would spell death. Adam had already experienced the first death of separation from God, but he would also experience a physical death, which is the king of terrors for all mankind.

This meant that man would have lived on forever in the garden, had he never committed sin. God declared that the punishment for sin is death **(Rom. 3:23)**, and there would be no exception to God's Law. When man dies, his flesh returns to the dust **(v. 19)** from whence it came. Every man that sins is placed under this curse of the Law of God, until he repents and accepts Jesus Christ's loving work of grace upon his life. Death is the ultimate curse of sin.

The Cure for Man's Plight

After God's announcement of His judgment to come upon His first family, He extended an offer of grace for their fallen and ruined lives. He offered to pick them up and wrap them in robes of righteousness, which were symbolized in the coats of animal skins. This of course, meant that most animals which had nothing to do with man's sin had to die to provide a covering for sinful man in the first blood covenant **(v. 21).** This picture was the prototype of the salvation that would be offered through the blood of Jesus to take care of man's sins. The blood shed by animals and their skins covering the sins depicted the ultimate cost of God's Son at Calvary.

There was only one creature had participated in the curse of sin and that was the reptile or serpent, so there would be animals used to bring forth a picture of God's plan. Lambs are one of the most defenseless animals that God created. They possess no ability to fight against their enemies, no power to climb for safety, nor any speed to run away. Jesus accepted the disposition of an innocent lamb by not trying to defend Himself, nor give an answer to His accusers at His mock trial. Isaiah stated His case clearly, *"He was lead as a lamb to the slaughter and as a sheep before its shearers is silent, so He opened not His mouth"* **(Isa. 53:7).** No doubt, this is why that Jesus is often referred to by the Apostle John as the Lamb of God and particularly in The Revelation.

Now, man would no longer possess the innocence of a lamb, but would now face a bitter struggle with a sinful nature and the consequences that accompany sin. The Lord God **(v. 22)** declared that through man's fall that his eyes were now open to the way of sin. He further stated that man would not be able to have both the tree of knowledge and the tree of life, since no one can have it both ways. Man had made his choice for knowledge by which he could gain upon his own. Therefore, the Lord God had to remove man from the paradise on earth, which contained the Eternal Tree of Life. From the beginning in paradise, sin has always carried consequences that are inescapable.

Man was sentenced to leave paradise and live in a much less productive world outside the garden, which was inhabited with thorns and thistles. The Lord God positioned angelic like servants called Cherubim at the entrance of Eden **(v. 24)** to keep any man from entering paradise until God's sacrifice was completed. His plan of redemption and paradise could be restored for those that would choose God's plan of redemption.

Chapter 9

The Curse on the Earth

1 Phillips, ***Exploring Genesis***, pg. 56.

CHAPTER 10

RAISING A LITTLE CAIN

Genesis 4:1–26

The verb "knew" means to know in an intimate and physical manner. Likely, Adam and Eve had sexual relations in the beautiful Garden of Eden where they had enjoyed their relationship spiritually and physically before sin entered into their lives. However, it was not until they were driven from the garden did Eve conceive a child. The Scriptures contend that they were naked and were comfortable with each other without any feelings of any improper behavior. No one can really say how long the first family lived in the paradise until they ruined their perfect habitat, but it could have been only months instead of years.

It seemed that Eve's rejoicing over a man-child **(v. 1)** that she viewed him as the redeemer that God had promised which would come from her seed to crush the head of the serpent. Eve was correct to believe that a redeemer would come through her seed; however, she was sadly wrong to believe that it would be Cain. This son would turn out to be a curse rather than a son of redemption that would produce great heartaches for his parents. It seems that shortly after the birth of Cain came the second son Abel, which means "vanity," although the telling of the events move much more rapidly than the actual happenings. Apparently the two brothers were somewhere close in age, yet they grew up in contrasting ways. While Abel became a herdsman and tended to livestock, Cain tilled the earth by farming vegetable crops from the ground that had been cursed with thorns and thistles **(v. 2).**

Years were passed over in a hurry as the young lads grew into adulthood **(v. 3),** which the writer describes as *"a process of time."* Cain brought forth an offering as did his brother Abel. No doubt, both men had learned from their father Adam, the practice of offering gifts to the Lord. Worship had been a significant aspect of the first family and became important to both of these young men. On one particular day, Cain brought the produce of his crops to offer God, while Abel came to his altar with an animal from his flock. The Lord honored the gift of Abel, but rejected the offering of Cain. Why would the Lord make such a distinction between the gifts of these two brothers? Some have suggested that the Lord received one and rejected the other because of the type of offering, while others have suggested that the quality of the offering was vastly different, while others have focused upon the attitudes of both men in their gifts.

The writer of the book of **Hebrews (11:4)** addresses the problem and declares that the each offering was a *"sacrifice"* type offering. The Levitical law describes many different and distinct types of offerings, such as the grain offering, burnt offering, peace offering, sin offering, priests offering, the first offerings, purification offering, sacrifice offering and many others common to the Jewish traditions. Once again, these men had learned from their parents about bringing offerings to the Lord. Since the offering was a type of sacrifice offering and maybe not a sin offering, which might give some insight as to why Cain's offering was unacceptable. It is clearly stated that Abel's offering was the *"firstling"* and that it contained "fat," which indicates a healthy animal. The Lord does require that his servants bring the best and always the first part. The statement concerning Cain's offering does not mention that he gave the first, nor the best, which leads many to conclude that Cain merely brought an offering from his produce or fruit. God is not well pleased with just any old offering, when he demands the best and the first part. Cain's gift may have been an offering, but it represented no real sacrifice.

Another question that must be answered, how did they realize that God had accepted an offering or not? In many Old Testament stories the Lord would consume the acceptable offering with a fire and smoke. The Lord commanded Abraham to prepare an offering of a calf, a goat, a ram, a turtledove and a pigeon **(Gen. 15:7–17)** and to split the animals upon an altar. Abraham drove away the birds from the slaughtered animals until sundown and then the Lord sent a flame and a smoking furnace to consume his offerings. If the fire fell from Heaven, then it consumed Abel's offering and not that of Cain. Cain became very angry at God and took out his frustrations upon his brother Abel. Likely, Cain was already jealous of his younger brother and now it really began to show up in his facial expressions.

Another conclusion that can be drawn from the two offerings and why one was unacceptable to the Lord is that the fruits and produce were not the right type of offering. In this suggestion, the Lord accepts only animals which shed their blood. It is true that there is no forgiveness of sin apart from the shedding of blood. The offering of Abel presents a picture of the blood of Christ at the cross dying as a sacrifice for sin by the giving of his own blood. If indeed, that Cain came with a trespass offering, then he should have brought a living animal to atone for his sin. An old gospel hymn asks the question, "What can take away my sin? Nothing but the blood of Jesus."

No one can say with certainty that Cain's offering was a sin offering; however, it is very obvious that Cain had sin in his heart when he came to the altar. The Lord would never have rejected his offering had there not been sin in his life. Whatever that sin was became even more serious as Cain became enraged with anger toward God. Unconfessed or un-repented sin always leads man to a deeper state of depravity and alienation from the Creator. So it is no wonder that the writer declared that Cain became *"very angry"* **(v. 5).**

The righteousness of a good man seems to provoke the heart of the unrighteous and such was the case of Cain as he turned his anger towards his brother, whom **Hebrews 11:4** says was righteous. Cain like his father Adam had now committed a serious offense against God that had alienated himself from the love of God. However, unlike his father, Cain does not acknowledge his sin and allow the

Lord to deal with his sin. There is not an attitude of repentance, but only an act of denial of any wrong. Much like in the case of Adam and Eve, God came with questions to confront Cain in his sin, yet he became very sarcastic about those questions. When the Lord asked him *"where is your brother"* **(v. 9)**, Cain said, *"am I responsible for my brother?"* Cain literally spurned God's offer of salvation, which would seal his own fate.

The Lord reminded Cain that if he did right then his gift would be accepted, but if he did wrong, then sin would be in the door of his heart **(v. 7).** The idea of the door would indicate that Cain had now opened his heart to sin just like his parents when they disobeyed the Lord in their garden home. The instruction was to deal with sin or else sin would dominate and rule his life forever. Cain would quickly become the slave to sin by allowing sin to remain inside his heart because sin is a tyrant that masters when is left to run free.

Jesus explained in the Sermon on the Mount that hate in a human heart is the germ for murder **(Matt. 6:21–22).** In this case the seed of anger quickly exploded into the first case of violence with human bloodshed in the Bible. Anger is like a stick of dynamite that can blow up in a moment when there is the least spark available. After the crime was committed, the Lord came calling upon Cain again, in order to bring him to confession of his wrong and to a place of repentance. However, Cain continued on his pathway of destruction by further denying any wrong doing. In fact, Cain lied to God and declared that he knew nothing about his brother whereabouts. With no repentant spirit, God announces that he is aware of the wrong, by declaring that Abel's blood had cried out from the ground **(v. 10).**

Notice the difference between man and the animal kingdom is demonstrated in the Word of God. Murder is an offense to God like any other sin, but nowhere does God forbid an animal not to kill another animal. If a lion kills another lion, no one declares that it "wrong," nor does anyone place the lion in jail **(1).** What one animal does to another animal is morally irrelevant. However, God forbids the act of murder among men, because man is completely different from an animal in the sight of God. Man's life is far more precious in the eyes of God. The Lord created masses of animals in one day that filled the earth, but He created only one man at a time. God keeps a genealogy of every individual that He creates.

A terrible curse **(v. 11)** would be placed upon Cain since he had declined to acknowledge any wrong doing or that any such act that God had described had ever occurred. The ground, which Cain used to grow his food for existence, would serve as a serious curse upon Cain's life. Notice that the soil had been cursed when Adam and Eve first committed sin, but an even worst curse was being turned loose to plague the soil that Cain touched. Blood would serve man in the act of redemption; however, when man rejects the blood of redemption, then that blood will become his curse. The curse for Cain's sin would mean that the resources of the soil would quickly be depleted wherever Cain tried to call his home, thus, making him a fugitive that would always be on the move. There would be no place where Cain could possibly settle down.

This homeless, wandering condition represents that of a lost or wayward soul with no relationship with the Lord God. Cain is no different from any lost man that has rejected the blood of Christ upon his life. He would be subject to a wasted life that would be filled with turmoil, instead of one with meaning and peace. God cannot allow the sinner to have inner peace until his sins are confessed and repented.

Then Cain turned towards the Lord to complain about his punishment that it was too stiff of a sentence that certainly did not fit the crime. He admitted that he would be unable to bear the weight of his sin, while the truth is that no man can bear the weight and the blight of sin. His sin brought fear to his heart that someone would surely do him harm for his sin. The question is raised as to who was out there that Cain believed would harm him? The Jews later were instructed by the Lord to take care of those that commit crime against their fellowman. In fact, the law of God specifically instructs the nearest of kin to find that person and to serve swift judgment upon the offender that suited the nature of the crime committed. In the case of murder, the nearest kinsmen would take the life of the murder, provided that it was determined that he intentionally took the person's life. Cain's own parents or his brothers and sisters would be responsible for taking his life **(5:4).** No other human beings existed at this time. Of course, God could have used an animal to take his life or some accident to befall him, without using a human being.

The Lord placed a mark upon Cain to protect him from those that would do him harm, thus the mark was a mark of mercy and protection rather than one for punishment. Instead of the Lord destroying Cain by removing him from creation, He left the door open for Cain to return and to repent of his sin. The Bible teaches that the Lord takes great pleasure in redeeming and restoring and gets no satisfaction in destroying people. The Bible never specifies what the mark was that was placed upon Cain. Much speculation has been given from that of a black man to a mark on his face or forehead that others could readily identify; however, the Bible really does not describe the mark.

If Cain had been made a black man, then the black race would have been wiped out during the flood. There is some indication from **Genesis 6:4**, that Cain could have become large in size and, perhaps, even a giant, which would explain the existence of such people in **Genesis 6**. If Cain became a large sized person, then others would fear him and not attempt to do him harm.

One of the saddest comments ever made about any man appears in this story, when the writer declares that Cain *"went out from the presence of the Lord"* **(v. 16).** No wise man would ever dare walk away from God. Cain did the unthinkable by turning his back upon the Lord and went to live as if God did not exist. Jonah makes a similar sinful mistake, as he determined to go to Tarshish, instead of going to Nineveh as God called him to share a message in that city. Consequently, Jonah bought a ticket that cost him much more than he wanted to pay with problems galore.

The Bible says that Jonah went down *"from the presence of the Lord"* **(Jonah 1:3).** There is no worse place in all the world to go than to go away from God, but Cain made this choice by denying God a correction of his wayward life. The very word Nod means to wander about, which means that Cain never had a permanent place to call home from the day that he walked away from the presence of God.

Cain became a nomad that would be forced to move from place to place.

Cain's wife conceived and bore him a son called Enoch. The verb "knew" **(v. 17)** again this indicates an intimate relationship between Cain and his wife was consummated. The often asked question at this point is "where did Cain obtain his wife?" Many contend that Cain found her in a new land; however the Bible is careful not to say that he found her in another land. According to the book of **Romans (5:12–19)**, all of mankind came from this one man Adam. Even Eve came from the one man Adam, which contradicts nature and the ordinary manner that babies are born in this world. The account in the Bible would bring students to accept the fact that the wife was one of his sisters. Adam and Eve had many daughters, although these are not commonly mentioned in the early genealogy of the Old Testament. In fact, not only are references rarely not given of daughters, but almost never are the names given of the daughters, except in special cases of a daughter story. Thus, Cain took a sister and moved away from his parents and other siblings that may have existed at the time of his departure. This has been the belief of many conservative early theologians, since the Bible account in **Romans 5:12** agrees that all mankind descended from this one man, Adam.

In the beginning of time, there were no problems with marrying sisters. In fact, this went on for some extended period of time as Abraham and others took their sisters or half-sisters to be their wives. In early civilization, there were no problems of marriages between close kinsmen with no known genetical inbreeding problems. Consider the fact in the perfect world of creation, there were no diseases in the early days of creation. Later the Lord forbid such marriages of near kinsmen, but at this time it was permissible, in order to populate the earth and even following the Great Flood.

Paul carefully explains that every man has the very same problem, which is a sin problem that brings forth the judgment of God upon the whole human race. Every man's heritage can be traced back to this first man named Adam. Thus, if every human being came from the loins of Adam and Eve, then the woman that Cain took with him when he left God's presence was a sister. The likelihood is that Cain was a grown man when he murdered his brother and generally the Hebrew was not considered ready for marriage until about forty years of age. If Cain was forty years old, then he likely already had several sisters. It is nowhere stated that Cain went out and found his wife in another land. In order for sin to be passed on biologically to every person, then it would certainly be reasonable that the wife had to have been an offspring of Adam and Eve.

The author is describing the pagan family that came from the loins of Cain and his family as they became progressively more wretched. Sin is often characterized in the Bible as dark and ugly, thus when people begin to entertain sinful practices, their lives become polluted like some cesspool. Cain's family took on this sin stained character moving further and further away from God in each succeeding generation. Cain began to build cities for his sons in hope that their fate would differ from his of roaming from place to place.

Lamech **(4:19)**, his name means "powerful," "conqueror" or "wild man" **(2)** likely became the first polygamist of the world as he took for himself two wives named Adah and Zillah. God's moral standards were being altered towards evil, as it has been said that Lamech had wives from A to Z (Adah and Zillah) **(4:19)**. Wife Adah bore his first son, Jabel, who became the leader of a tribe of people that were shepherds living in tents. However, his brother Jubal chose another occupation of a manufacturer of musical instruments **(4:21).** The first two sons of wife Adah were very different as were Cain and Abel, which without a doubt, created sibling rivalry.

Wife Z bore Lamech a son called Tubal-Cain, a teacher and craftsman with metals such as bronze and iron. Tubal-Cain is the last named son born to the descendants of Cain, which was Cain's great-great-great-great-grandson.

Lamech, like many wicked men, began to boast of his love affair with evil. He proudly declared to his wives A and Z about how his wickedness had far exceeded that of all his ancestors. With a sinful, prideful, and arrogant heart, Lamech declared himself a murderer of a man that had wounded him. The Bible is silent on the circumstances of this conflict, except that Lamech was somehow wounded in the scuffle, before he landed the final blow to the man. The victim very likely may have been one of his own kinsmen. He had taken the opposite course of action that the Bible teaches, as it declares that *"vengeance is mine, thus saith the Lord"* **(Deut. 32:35)**; however, Lamech took the law into his own hands and slew the man.

Lamech further warned that any man who messed with him to bring him hurt, that he would receive the wrath of Lamech at seventy times greater than that of Cain, his great-great-great-grandfather. This advanced form of evil strikes a serious contrast to the Lord's teachings on love and forgiveness in **Matthew 21.** Lamech wanted people to be advised of their destiny if they crossed him or angered him in any manner. Jesus declared that anger is the germ or root cause of murder in the Sermon on the Mount **(Matt. 5:21–26).** Like many men today, Lamech was proud to be known as a man of much anger and saw it as some noble quality of his character.

The genealogy of the wicked family of Cain ends with unregenerate Lamech, who seemed to be the worst of the worst. This family of nomadic people filled their lives with evil and appears to never have come into the presence of the Holy, Righteous God that had created them. They wondered from place to place outside the will of God becoming more and more corrupted.

The writer now shifts his message back to the family of Adam and Eve. Again, the Hebrew verb *"knew"* **(4:25)**, which means "to love sexually or intimately" is used by the writer to share the story about the substitute son of Adam and Eve named Seth. Seth became the appointed seed to replace Abel, whose life was snuffed out by his brother. This new account brings forth the family genealogy of the people from Adam that followed the Lord God.

The people would be those that the Lord would be able to use for His family heritage. He declared *"men began to call upon the name of the Lord"* **(4:26).** This sounds much like **Romans 10:13**, when Paul said, *"Whosoever shall call upon the name of the Lord shall be saved."* These people began to express their need and faith in the God of creation. Seth and his son, Enoch and their offspring would be the Godly lineage that obeyed the Lord God, which stands in a sharp contrast to the family of Cain that totally disregarded God.

Adam, Eve, Seth, Enoch, and their families recognized that the Lord was important to their lives. Chapter 5 of Genesis is the story of Adam and his descendants through the substitute son, Seth. The story begins with Adam, which was the only man that was created in the expressed image of God. Adam was born unlike any man, since he was born sinless and not shapened in sin and iniquity without any earthly parents.

However, after Adam and Eve sinned they passed on to every baby the sin nature and thus, every child has inherited sin and iniquity. The story in Genesis five begins with a retelling of the creation account where God had created them male and female for compatibility. He did not create man for man, nor woman for woman as many would have man to believe in the present generation. God did not create Adam and Steve, but Adam and Eve. God abhors homosexuality, lesbianism, and transgender relations that present world seems to have gone soft on today. God has fully approved sexuality between one man and one woman bound by the laws of marriage. This is the only accepted practice for man to pursue in the sight of Holy God.

Leviticus and Romans both make God's position on human sexuality perfect clear. God declared to man, *"thou shall not lie with a male as with a woman. It is an abomination"* **(Lev. 18:22).** Again in **Romans 1:27**, *"Likewise also the men, leaving the natural use of the woman, burned in their lust for one another, men with men committing what is shameful, and receiving in themselves the penalty of their error which was due."* Paul goes on in the following verse defining the situation further that these men did not like to retain the teaching of God in their lives. These passages of Scripture are not ambiguous and make God's position on marriage from the beginning for one man and only one woman. Like the family of Cain, men have decided to ignore the teaching of their Creator and choose their own standards. Man is free to do so, but he should be prepared to suffer the consequences of his sins when he makes these defiant choices.

From the day that Adam was born as a full grown man until his death, Adam lived 930 actual years. Adam lived 130 years before he begot Seth, his third born son. At that point, the Lord injects another interesting note regarding Seth. Seth is born in the image of Adam **(5:3)** and not in the image of God as was stated in the birth of Adam. Adam lived long enough to father several sons and daughters **(5:4),** which was important to populate the earth in the early generations. This helps the Bible student to better understand where Cain got his wife before going away from God. Some who try to deny the literal interpretation of the Scripture at this point suggest that these numbers represent months rather than actual years. That view presents some major conflicts, since Enoch **(v. 21)** would have been only sixty-five months old or five and a half years old when his first son was born. When symbolism is used, it always follows after some great truths that have previously been established; however, in this account, there are no facts that precede the creation story.

Another interesting note is added to the genealogy of each of the descendants after the years of their lives is given and that each one died. This statement stands in opposition to the words of Satan to Eve in the garden, when he stated, *"You surely shall not die."* God's record proves Satan is a liar, since each descendent ultimately died as God had declared. In **Hebrews (9:27),** *"And it is appointed for men to die once, but after this comes the judgment."*

The Promised Seed that God revealed to Adam would come through the descendants of Seth, thus the writer gives a complete list of the first born son of each and their length of life. This Bible record of years proves to be an invaluable aid in determining the years prior to the flood and ultimately the number of years since creation. There arose one from Seth's descendants that stood out from all the rest of his kinsmen, which was named Enoch. He was a special Godly man that God chose to carry on the human race.

Enoch represented the sixth generation from Adam and the Bible states that this man, perhaps, like no other man of any period of time walked three hundred years with the Lord. The verb *"walked"* closely indicates that Enoch was in step with God, as well as, in motion for the Lord. Also, worth noting is the fact that God has no feet as does man, since He is a Spirit Being. The writer is simply describing their relationship in anthropomorphical terms or human terms, since man cannot describe the appearance of the Lord in His spiritual nature. The bottom line at this point is that Enoch and His Lord had a beautiful, satisfying and meaningful fellowship.

The Genesis writer on two occasions, describes a progressive relationship that was going somewhere by the use of the term "walked with God." After many times of walking and fellowshipping together, God took Enoch on to Heaven before he could live out the usual life expectancy of his forefathers. Enoch only lived 365 years as compared to 900 years plus for most of his kinsmen. It would certainly be safe to state that Enoch loved God's company far more than he loved the things of earth, so the Lord transported Enoch on to heaven. Every single person born died, except for Enoch that God took home with him **(5:24)** and Elijah, who God took with a chariot of fire **(2 Kings 2:11).**

In chapter 5 of Genesis is absolutely important to Bible students to make the connection of the actual time frame of events that transpired in the early generations of mankind. Chapter 5 stands as the antithesis of the devil's statements to Eve when he was persuading her to sin in Genesis 3. The serpent said, *"you will not surely die"* **(3:4).** However, God clearly proclaims the opposite as the Bible states their age and then their death. Someone said that Enoch regularly walked with God every day. One day, Enoch just kept on walking with God, until late in the day and God said to Enoch, "We are closer to my house than yours, so come on home with me." That may have been the case, since Enoch was such a man, unlike some others that claimed they were after the heart of the Lord.

Genesis 5 also gives us the time from the creation of Adam to the coming of the Great Flood. When the fourteen generations are added together, there is a period of 1,656 years **(2344 BC)** from Adam to the Flood. Without chapter 5, it would be hard, if not impossible to credit the period of time the earth was spoken into existence. This chapter informs all generations that we are merely one generation away from becoming pagans. Should one generation fail to teach their children about the Lord God, then it will produce a godless society that holds no respect for God. The world gives its version that the earth is millions of year old and that man finally evolved from some lower species of animal life around two hundred thousand to five hundred thousand years ago. An article appearing in ***National Geographic*** **in January 2013 issue** declaring that man lived for some two hundred thousand years in Africa before he began to migrate into other parts of the world about fifty thousand years ago. These stories are widely published in textbooks, which deny the story of the Bible **(3)**.

Additionally, Genesis 5 tells us that the Methuselah, the oldest man who ever lived that he lived 969 years and then died. Methuselah was the grandfather of Noah; however when the years are added up he died in 1,656th year of history, which was the very year of the Great Flood. He most likely died in the flood in his wicked condition without the Lord. Was Methuselah one of those trying to get into the ark, when the flood waters began to rage? Surely Noah would have let him inside, but God had shut the door and there was no person that could open that door. This is the same situation at the coming of the Lord Jesus, when He comes for the Rapture of the Church. People will be repenting and wanting to get into the ark of safety and shelter; however there be no more time to make any decisions.

Chapter 10

1 Lisle, Jason, ***Why Genesis Matters***, Institute for Christian Research, Dallas, Texas: pg. 13.

2 **Phillips, *Exploring Genesis*, pg. 73.**

3 Stringer, Chris, ***National Geographic***, Global Journey, Natural History Museum, London: January 2013, pg. 48–49.

CHAPTER 11

HOW DID THE FLOOD OF THE AGES TRANSFORM THE ORIGINAL CREATION?

Nothing has ever changed nor transformed the surface of the earth like the worldwide flood described in **Genesis 6–8**. Like sin that changes everything and every relationship in the life of an individual, the earth was drastically transformed by the ugly curse of sin on the face of the earth. In this chapter, the discussion will be centered upon how all the works of creation were reformed and renewed at the very same time that it was being destroyed. Here are a couple of the questions that people frequently ask which deal with God's judgment during the Great Flood in the day of Noah. Is there any scientific evidence to support a worldwide flood over it being a regional flood? Why did God bring forth a destructive flood and kill so many people at one time while destroying the entire world?

In the midst of God's wrath, He always holds out a lifeline to any and all who are repentant that look to Him for grace. God informed Noah, a righteous man what He was about to do to all the earth and asked Noah to prepare his family for the consequences of the impending judgment of God's wrath. The Bible declares that "Noah walked with God" **(Gen. 6:9),** which is the second person that such a statement is made in the Bible. The first was Enoch in

Genesis 5:24, as it declares Enoch walked with God." Not only did these two men believe God and obey Him, but walking indicates both spent considerable time in the presence of the Lord God. Walk can mean the same as to live with God. While most of the world had forsaken the Lord, Noah remained faithful to the Creator.

Noah was a God fearing man that truly believed God, began a long process of making the preparations that God had instructed him to perform. The Bible says that *"Noah did according to all that the Lord commanded him"* **(Gen. 7:5).** This means that Noah did not leave out any of the instructions of the Lord until the ark was completed. Noah diligently worked on the ark for years with his three sons and perhaps, Mrs. Noah. Some have suggested that Noah may have hired people to help him complete the work of the ark; however, the Bible is silent of any such labor being performed other than Noah's family.

Remember that it had never rained when God told Noah to build this huge floating crate for all the animals and his family. It was truly a vessel for safety and survival and not a ship with a rudder, since God would navigate this vessel. Image the ridicule that Noah received from all the people, the laughs and mockery. It appeared that nobody believed God, except Noah and his family, thus they were all disbelievers. *"God saw that the wickedness of man was great in the earth, and that every intent and the thoughts of his heart was only evil continually"* **(6:5).** It is hard to believe that the whole earth fell into such depravity, except for the family of Noah.

It had to have been hard to bear the persecution that Noah and his family endured for those days of construction of the huge boat, when it had never rained before. God had always supplied a cool mist of water to rise up from the ground that kept the earth well watered like some modern day irrigation system. His irrigation system made things pleasant until the sins of wickedness pushed God to bring forth His judgment. His plan would destroy everything and everyone, except those on board the ark. This meant that God's judgment would affect every inch of the earth with this catastrophic event, which would bring the destruction to everything on earth, while sparing all inside the ark. The next world destruction will not be with water, but will be a consuming fire of the wrath of God. Again, God has prepared salvation for those that have chosen God's way, thus those will be inside His ark of safety and spared from another harsh judgment in the End Time.

There is little doubt that Noah entertained many questions about why he was building such a huge boat from his neighbors. They knew that Noah was an honorable man and likely gave some considerations to his sobering message. Human nature generally puts off any serious decisions, until they begin to feel the wrath of the Lord. When that happened in this story, it was now too late for anyone to make a decision or to heed the warning of destruction. The same type of situation will come again when Christ returns in the End Time.

One common thought from textbook teaching that the Great Flood was a regional flood and not a global flood. Almost no teacher would attempt to give any acknowledgement of a regional flood after reading the awesome flood described in Genesis. If this were a regional flood, then why did God tell Noah to build such a boat, rather than just telling him to move a few hundred miles away from home, where he might be spared. A big argument for the universality of the flood comes in Genesis 7:19–20, when the Bible says, *"The waters prevailed exceedingly on the earth, and all the high hills under the whole heaven were covered. The waters prevailed fifteen cubits upward, and the mountains were covered."* This clearly states that the highest mountains were covered by twenty-two feet of water, thus there is no way that this was a local or regional flood. Another good question that people often ask, did God destroy all the people and animals outside of the ark. Yes, the Bible says that He destroyed everything that had breath died in the flood, which would include all those land animals. There was a possibility that some of the sea creatures may have survived in the treacherous waters outside the ark.

The Bible describes the term of *"the windows of heaven being opened,"* **(Gen.7:11)** which is extremely important for the understanding of the extent of this great flood that covered the entire planet of earth. The CSH version declares, *"The floodgates of the sky were opened."* Whether one thinks of gates, windows, or whatever, it was far greater than little leaks in the canopy, but rather a total collapse of the canopy that God had initially designed to last for all eternity for the perseverance of a pleasant life under the tent of His protection. Man's sin had become so offensive before God that He had opened a massive pathway of destruction upon His beautiful perfect planet earth, as well as, every creature that inhabited the earth, except those on board the ark. Only Noah and his family received the grace of God that protected them through the judgment of a Holy God. Just the weight of the water from above was capable of unbelievable destruction apart from the ruptures in the crust of the earth.

Jesus said in Matthew that all life outside the ark was destroyed. He declared, *"They did not know until the flood came and took them all away, so also will the coming of the Son of Man be"* **(Matt. 24:39).** If Jesus intended His statement to mean a partial judgment, then He would have meant that the next judgment by fire would also be a partial judgment **(1).** A further question that others ask, is there scientific evidence for a global flood and indeed there are books filled with hard evidence all around the world. However, much of the world has explained it away, but it is obvious to the observing student of science and history that the proof abounds that supports the global flood described in Genesis. Many aspects of this evidence will be discussed in the forthcoming comments. A student of science would profit much by studying the related Scripture in Genesis 6:1–11:9, which includes the Tower of Babel.

Another question is "why did God do what He did" in killing so many people and animals with an international calamity? That question is answered in the Bible **(Gen. 6:5–6).** God clearly states His case for destruction of the earth and all of its inhabitants, except for Noah and his family. The world's population explosion led to spiritual decay, to shameless depravity, to social dilemma, and to strong delusion of the truth. The runaway population growth led to a great apostasy and forbidden sexual liaison between women and fallen angelic beings producing a hybrid race of giants **(2).** The Lord God is not a cruel Being, but is full of mercy and grace and is longsuffering with all mankind. Sometimes man crosses a line of no return and there is stated in the Bible a sin unto death **(1 Jn. 5:16).** According to the Apostle Peter **(2 Peter 2:5–7).** the Lord left *"an example to those coming afterwards that would live ungodly."* The Bible declares that God takes no pleasure in destroying, but rather in redeeming sinful men that will turn to Jesus.

The earth today cries out with mountains of evidence on the earth's surface of a tremendous flood. There are craters, canyons, coal beds with layers of the earth's crust signifying the results of a major global flood **(3)**. If a student of geology would seriously observe the Grand Canyon in the western United States, he would have a wow moment. There lies the biggest ditch in the world carved out of solid rock by water. People know that rain in that region of the planet is never furious enough there to cut through that type of rock. Most science folks would explain that it took millions of years of erosion to cut the canyon from such rock. Why then is there not any evidence that the river at the bottom of the canyon not still cutting down the rock layers over the past several hundred years? It had to be a massive amount of rushing water with tremendous force that formed the canyon, since there is no other scientific way for such a thing to occur. True science says one of two things produces true scientific facts. One is that someone must observe it happening or two it has to be proven in a scientific laboratory in a controlled environment. Neither of these are true with most so-called science experts. Science experts have been totally unable to produce any true facts or show through scientific experiments that such a canyon could be formed slowly.

Finally, the floating boat was complete with all of the slime and pitch to make it waterproof. During that time, the Lord God spoke to Noah and said, "Come into the ark" (7:1). This command to "come" would clearly signify that God was already inside the ark. Come in the ark and the same as being "in Christ." Because Noah believed God, he went in with all the animals with him that God brought to the ark. Some people contend that Noah must have had to have a gigantic animal rodeo roundup. It did not happen that way, since the Lord brought the animals to Noah, rather than his rounding up all those that went on board. The Lord God was in charge and

He had a well laid out plan of the specific animals and size that He desired inside the ark for the preservation of each species.

Perhaps, there were many sea living creatures and even many of the large sea mammals that could have been able to survive in the treacherous waters for over a year. Nowhere does it say in the Bible that Noah had to have all the grown up animals, so many of them could have been baby elephants or baby dinosaurs. The Bible does say that some animals were 2 of a kind, while others were seven of one kind for those clean animals **(Gen. 7:2).** The extra clean animals were in anticipation of the new rule, after the flood, that man should henceforth eat meat **(4).** It appears that Noah, his family and all the animals had to sit and wait inside the ark for seven more days before the rain began to fall upon the earth. After the seven days of no rain, then the rains began to fall from the sky. It was the seventeenth day of the second month and in the six hundredth year of Noah's life that all the fountains of the great deep were broken up, and the windows of heaven were opened **(Gen. 7:11).**

These two major catastrophic events were happening together that would reshape the entire earth and had worldwide effects upon the climate of the entire universe. The earth was in harmony with all other aspects of the universe in its perfect foundation that God had created. As a result of God's wrath, all of the planets would undergo a gigantic cosmic makeover, but especially the earth.

When the great deep **(Gen. 7:11)** broke up allowing the flood waters from the belly of the earth, began to gush from the earth leaving rock fissures in the new earth's crust. These fractures in the earth continue to create problems as the rock layers shift causing earthquakes and other geological problems from deep within the earth. Further fractures likely occurred by the heavy weight of the flood waters as they pushed the lowlands down and caused the mountainous area to rise higher and higher. This upheaval of the highlands also created tremendous runoff of the flood waters causing canyons and huge waterfalls like the Grand Canyon and Niagara Falls during a period of a few days rather than billions of years. The proponents of the billion-year theory suggest that there was no way for such serious erosion to take place over a short span of time. All of these discount an all-powerful Supernatural God and a worldwide flood, such as the biblical flood described in Genesis 6–8.

The physical fractures in the earth's crust are found today around the globe. The result of one of these fractures is the New Madrid fault line that runs up and down the Mississippi River through Missouri, Arkansas, Illinois, Kentucky, and Tennessee. On December 16, 1811, one of the most powerful earthquake's in American history caused the Mississippi River to run backwards forming Reelfoot Lake in Northwestern part of Tennessee. Over a period of a few months from December 16, 1811, into the early months of 1812, there were over 1,800 earthquakes around New Madrid, Missouri (Boot heel of Missouri), which helped to form Reelfoot Lake.

Those earthquakes were felt as far away as Washington, DC, and Charleston, SC. The quakes reshaped the northwest corner of Tennessee around Tiptonville causing the uplift of the land, which is now called the Tiptonville Dome. At the same time of the shift, a swampy parcel of land, approximately fifteen thousand acres sank several feet forming a huge lake called Reelfoot Lake, which today is a state park area. A museum in New Madrid, Missouri, was founded several years ago that describes this geological phenomenon.

Experienced river boat pilots that were guiding their flatboats down river, were greatly surprised on December 16, 1811, when the river swept them back up the river for more than a mile. The river continued running backwards for many days until all the lowlands were filled with river water forming this new lake. Several islands along the river completely disappeared from the landscape during this same time frame. Many residents reported strong sulfur smells covering the entire area for weeks with a cloudy haze of gases erupting from the surface of the earth. Trees along the river were blown out of the river from the powerful forces of nature.

The series of earthquakes around the New Madrid fault line created big problems for the residents in the early 1800s, but small quakes and tremors are still occurring along this fault line at present. Other fault lines are present throughout the surface of the earth. A large shift on a fault line occurred in the ocean off the coast of Indonesia in December of 2004, causing a huge tsunami that destroyed nearly half million lives as it swept away practically everything in its path in at least a half dozen nations in the Pacific Rim of Asia. Again in February of 2011, there was another such episode with an earthquake off the coast of Japan, which created massive devastation and loss of life from the tsunami effect. In California, there is another large fault line that is a continual threat to the citizens of that area called the San Andreas Fault. There is a continuous amount of earthquakes and tremors that occur along most of these fault lines, which run through most of the western coast side of California. Perhaps, this is what the Psalmist, was proclaiming when he declared, *"Then the channels of the waters were seen, and the foundations of the world were uncovered at His rebuke"* **(Ps. 18:15).** It is true that fractures exist all over the earth and in the ocean floors that are referred as fault lines. Many of these have been at the heart of the tsunamis described above and many other times during the past four thousand to five thousand years following the Great Flood in the days of Noah.

It certainly appears that these fault lines are a result of the Great Flood in the day of Noah. The lowlands received tons and tons of water which pressed the earth downward, while at the same time it was lifting up the hills and mountains making them higher than the pre-flood days. Truly, the landscape of the earth was totally transformed by the Great Flood. Many scientists declare prior to the Flood, the surface of the earth was about 30 percent water and the land being approximately 70 percent. After the flood, that formula was reversed as today the surface of the earth is approximately 70 percent water and 30 percent land. Perhaps, this helps to explain what happened to all the water that fell from above when the windows of heaven were opened **(Gen. 7:11)** and the canopy of water vapor fell upon the earth for forty days and nights. All of these waters, plus the waters that spewed from the belly of the earth actually raised the level of the ocean water significantly from those that existed prior to the flood. On the ocean floors around the world are land bridges that connected the land areas together prior to the Great Flood. These land bridges became submerged in the oceans after all of the extra water from the forty days of serious flooding.

Again, just exactly what was the Genesis writer speaking about, when he said *"the windows of heaven were opened in **Genesis** 7:11"*? When God was creating the heavens and upper atmosphere on day 2 of creation, *He divided the waters above the firmament and below the firmament* **(Gen. 1:6–7).** On that very same day the Lord God created a water canopy or likely a water vapor barrier completely encircling the sphere of the earth in the upper atmosphere to protect the earth from any harmful radiation rays entering into the earth. The Psalmist said it clearly, *"His canopy around Him was dark waters and thick clouds of the skies"* **(Ps. 18:11).** Many scientists reject this whole concept of a water vapor canopy, while the Lord places these features in His Word for all to see. This barrier would prove to be essential to protect all the creatures from the exposures to harmful radiation while providing a very pleasant atmosphere for all of God's creatures. The Creator had a working plan in hand prior to His creating His Kingdom, plus protection for all mankind.

This barrier produced a greenhouse type of perfect environment surrounding the entire planet and giving it a pleasant 72-degree comfort setting all over the earth's crust from the north pole to the south pole. Remember God made a perfect place in His divine creation, but His judgment fell and destroyed that blessed setting by removing many of God's blessings for man's comfort because of man's sinfulness. In the case of such gross sin, God would remove many of His barriers of protection for man.

Once the water vapor canopy was broken open that water stored above the surface of the upper atmosphere began to gush down in great torrents. Some scientists have suggested for the water to gush through the windows of heaven with the force described in the Bible that that canopy of water must have consisted of about forty miles of vapor to have rained continuously for forty days and nights. This was an enormous amount of water, which quickly produced a horrifying deluge of water that spared no one or nothing in its path.

The constant torrents of water crashing onto the earth, plus the water gushing from underneath the earth's surface created the perfect storm of flooding. Within the first few days the earth was submerged under water, yet the water continued to come in the same manner for forty continuous days, according to the Scriptures. No doubt, God protected the ears and eyes of the eight people on board the ark from much of what was happening on the outside of their ark of safety. Noah believed God and followed all the instructions that the Lord gave him, so he was resting in the arms of God.

The volume of water that came from both directions was a huge and tragic event for all of life. Not only were the people frightened, but likely so were the animals, since none of these had ever seen it rain upon the earth. God had provided an irrigation system to rise up and water the whole surface of the earth **(Gen. 2:6).** This was a highly sophisticated irrigation system that far out classes anyone's system in the technical world today. God's system never required any maintenance, but efficiently worked for 1,656 years from creation. All things were transformed because of the sin of the people which inhabited the earth and now God was pouring out His wrath upon the earth and all of its inhabitants. *"The Lord sat enthroned at the Flood and He sits as the King forever"* **(Ps. 29:10).** God orchestrated the Flood of all times to express His displeasure over the horrible high level of sin in the heart of mankind.

Sin carries a curse that follows afterwards to bring destruction. In the New Testament, the Bible says **(Rom. 6:23)** that the wages of sin is death. Because God is Holy, then He must deal with sin and He will not postpone His judgment forever when sin runs wild. There will certainly be a great judgment of the Lord upon the earth in the end of times, but there is always an ongoing judgment of the Lord every day the world lives in stubborn rebellion to the laws of God. All will suffer the consequences of disobedience to God's commands.

Once the forty days of continual flooding passed, there would come an even greater transformation of the landscape. There was an additional 110 days when the waters were still abounded with great force, before they began to go down. When the skies cleared and later the mountaintops began to appear, the waters were rushing at such a rapid force that the weight of the water in the low places pushed the earth down. At the same time that the low places were sinking lower, the mountains were being pushed upward causing them to become higher and higher. This upheaval of the land created much taller mountains than those prior to the flood, while cutting out new and larger rivers all over the surface of the earth. Many of the land bridges that connected the earth prior to the flood had now disappeared as the ocean levels had tremendously been elevated. Most of these natural land bridges before the flood are still found in the ocean floors buried hundreds of feet below the surface of the water.

Another significant change that took place was produced by the sudden temperature change all around the earth, especially in the polar regions. These areas began to freeze and developed a huge ice age, freezing the bodies of animals that been living in these regions at the time of the Great Flood. In fact many of these creatures have been dug out of the frozen tundra with green grass in their mouths and stomachs. This is another piece of evidence for the universality of the flood. The animals were pleasantly grazing along in the green pastures in the polar areas of the earth, when the flood water began to overtake them. There was no place to run away and hide from this flood. There are estimates as high as 5 million mammoths, whose remains are buried all along the northern Siberia and into Alaska **(5).** No longer would these regions of the earth enjoy the pleasant 72 degrees and remain the home for so many of God's creatures. There are a few creatures that have been able to adapt to the harsh, frigid temperatures of the North and South Poles today.

All around the earth, the remnants of the flood's runoff waters have left plenty of evidence the paths that it took to finally look at the ocean floors. There remain thick deposits of alluvial sediments from every body of land extending out into the ocean floor.

One of the major pieces of evidence for a worldwide flood such as the Genesis Flood are the fossils records all over the earth from the mountaintops throughout every ocean floor. Trillions of fossils of sea creatures in the high up mountain ranges should prove that there was a flood at some point of time. Most of those that oppose the Bible records, especially the Genesis stories attempt to prove their theory by adding a few million or billions of years for the age of the earth. Those people generally believe in a local flood because they have accepted the highly believed evolutionary history of the earth, which interprets fossil layers as the history of the sequential appearance of life over millions of years **(6).**

Modern scientists understand that these fossil records once were the greatest evidence for a global flood that had been found buried in sediments of mud sand and rock layers around the earth. Thus, after adding the millions or billions of years they persuaded their followers that all of the sediments formed over the long period of time rather than a sudden event that buried all of the fossils. Unfortunately, many Believers in the Bible account have bought into the theory of millions or billions of years, which has no legs to stand upon.

The graveyards of thousands of different animal fossils have been discovered mixed in various locations around the earth. Some of these sites hold the fossils and bones of animals that have long been extinct. There are great mammal beds of the Rockies; the dinosaur beds of the Black Hills and the Rockies, as well as, in the Gobi Desert; astounding fish beds of the Scottish Devonian strata **(7).** These huge graveyards cannot be ignored that hold great evidence for a global flood is the only answer to explain such strange discoveries. What would be the chances of all these animals dying about the same time at the same place, unless they were hosting an animal family reunion when they all suddenly died at the same time.

Furthermore, there exist large deposits of rock, skeletal parts of animals, aquatic life and fossils scattered all over the earth that gives evidence of a global flood that left its tracts in the landscape. There exist fossil records in most any place, whether on land or in the sea that suggests a global flood once happened. There are massive coal deposits left all over the earth from the sediments of plants and fossils from the flood. People widely accepted the global flood story from the day of Noah as fact until the recent years. Once the unbelievers of a Supreme Creator began to write the science textbooks, there was very little to question about the Great Flood. This story has been passed along to every generation.

One famous archeologist, Robert Ballard recently discovered evidence the supports a super flood in the region of the Black Sea. Ballard found the remains of the sunken Titanic in 1985. He claims that at the bottom of the Black Sea is plenty of evidence that this once fresh water lake was suddenly changed by an enormous wall of water from the Mediterranean, which was two hundred times more powerful than Niagara Falls, swept it and everything else away. Ballard suggests that the small lake became a large sea after the flood. He and his team of scientists have discovered evidence four hundred feet below the shoreline that suggest a huge flood swept from Europe into the Mideast Asia area. Ballard does not believe in a worldwide flood, but more of regional flood. He estimates from his findings that it happened nearly five thousand years ago. Ballard contends that there are ancient villages that have been buried in many places from some kind of huge flood, which lie covered underneath water **(8).** The truth is that there is plenty evidence all around the world where enormous evidence exist for a global flood such as the Grand Canyon. All of this evidence supports the biblical story of a huge worldwide flood in the day of Noah.

After the flood waters abated and the land was dry, God kept Noah and his family onboard the ark until all the threats of decomposing bodies were gone. He did not want Noah and his family to be faced with a death in his family until they had an opportunity to begin the repopulation of the universe. Noah and his family were actually on the ark for just over a year, with the addition of the days of Genesis 7–8 add up to approximately 378 days. Perhaps, all readers need a reminder that the actual days of the rainfall only lasted 40 days and nights of those 378 days inside of the ark. It is important to note that, we understand that every person on planet earth today is a descendent of Noah. It one believes the biblical account, there is no other people left on earth, except this one family after the flood. Thus, all come through this family of Noah. God started again to populate the earth with His chosen people. At the same time, God promised that He would never bring such a destruction upon the earth by water and placed His rainbow in the sky **(9:13)** for all to see that His promise is still good today.

However, God did not promise that He would not destroy the earth again. In fact, He gave a promise that the next time that it will be by fire, instead of water. In **2 Peter 3:10**, the Lord said through Peter, *"The day of the Lord will come as a thief in the night, in which* *the heavens will pass away with a great noise, and the elements will melt with a fervent heat, both the earth and the works that are in it will be burned up."* Again, nothing will be left of this earth and no one will survive, unless a person is on the ark of safety, where the Lord Jesus is on board. No Jehovah Witness can remain on earth and be spared of this burning inferno. Over the years, this author has talked with many of these followers of Jehovah, who contend that this will absolutely not take place. The truth is that God will only allow wickedness to go so far, before He brings down His judgment and it will happen again according to the Bible.

There are also the scoffers who are present in every generation, who ridicule the people of the Lord and ask sharp questions about God. Apparently, these will multiply in the end of time, as they walk in their own lusts **(2 Peter 3:3).** But Peter declares about another group of people that will be present and those who are willfully ignorant **(3:5).** Those people are either ignorant on purpose or they want to forget about God's promises.

One amazing piece of scientific evidence that suggests a worldwide flood is the fact that fish fossils exists all around the earth in rock formations. Under normal conditions, there are little or no fish fossils, since these creatures are eaten by other aquatic creatures long before they can be fossilized. But the earth is filled with fish fossils in most layers of rock that suggest the Great Flood was indeed worldwide. During this same period, which was approximately about 4,400 years ago all of the large size or enormous animals seem to have disappeared about the same time as the Flood. This would be consistent with the mammoth size animals that were found around the earth, even on the poplar caps frozen whole with green grass in their mouths.

The water vapor canopy that produced a greenhouse type of effect around the globe produced a pleasant temperature of around 72 degrees all over the surface of the earth. Remember when God called all things into existence during His six days of creation, He made the animals in abundance and filled up the earth with each species. It was only with the creation of mankind that God originally made only one person, which was Adam. The animals on earth, the birds filled the airways, the aquatic creatures filled the seas as the larger animals that lived upon the earth filled up the earth from one end to the other end.

The purpose of this study is to educate and prepare God's people to KNOW the truth of God's holy Word and be able to recognize the error that is being widely offered in the world today. Additionally, this study is intended to give students of the Word truths that will enable them to give an answer for those that seek to deceive or mislead others erroneously away from the Lord Jesus. May the Lord bless those students of the Word of Truth to be able to give answers from the Holy Bible.

Chapter 11

How the Flood Transformed the Earth

1. Ham, ***The Answer Book***, Vol. 1. pg. 137.
2. Phillips, ***Exploring Genesis***, pg. 79.
3. Ham, ***The Answers Book***, Vol. 1. pg. 138.
4. ***Exploring Genesis***, pg. 85.
5. Morris, Henry, and Miller, William J., ***The Genesis Flood***, The Introduction to Historical Geology, 1952, pg. 156.
6. Ham, Ken, General Editor, ***The New Answers Book 1***, Answers in Genesis, Master Books, Green Forrest, Ark: 2011, pg. 136.
7. Morris, ***The Genesis Flood***, pg. 161.
8. ***AFA Journal***, The American Family Association, from Fox News, www.foxnews.com, 12/12/12, March 2013 issue, pg. 6.

CHAPTER 12

THE AGE OF THE EARTH

There is a wide spread range of years debate from a few thousand years to over twenty billion years. How does a serious Bible student decide which is true? Science often gives several answers, while conservative Bible teachers give a far different view. The biblical record dates the earth as slightly over six thousand years in age from creation to the present time and not the millions or billions of years that are often tossed about as an accurate figure by science. Even some Christians have adopted the billions of years theory without a question of any evidence. The biblical record can be validated from the dates and years given in the Bible story. For example, in Genesis 5, the length of years in the era from Adam to the Great Flood are easy to total. A count of the Genesis record reveals that there are 1,656 years that exist from the birth of Adam until the coming of the Great Flood. The Bible record gives a period of approximately 2,400 years following the Flood until the coming of the Christ Child in the New Testament.

The dates from the end of the flood to Abraham and his son Isaac and Jacob are easy to trace. Jacob's and his family's travail in Egypt was four hundred years and the forty years of wilderness wanderings are dated. The lives of Moses, Joshua, through the Judges and the Kings can all be established and documented.

The world of science has an entirely different version that dates the earth over twenty billion years and beyond. Their belief is a religion without God. Kent Hovind, noted scientist, author, and advocate speaker for the biblical view is just one of many scientists that is an exception to the rule. He makes this statement in one of his writings to explain how far off base the scientific world has gone to prove evolution. "In children's fairy tales, we are told FROG + MAGIC SPELL (usually a kiss) = PRINCE. In modern science textbooks, we are told that FROG + TIME = PRINCE. The same basic fairy tale is being promoted in textbooks today with the new magic potion being cited as TIME. When the theory of evolution is discussed, TIME is the panacea for all the thousands of problems that arise. Time is the evolutionists' god" **(1).**

There are several creation writers today that researched for thousands and thousands of hours this old creation story and have produced libraries filled with credible writings that refute the theory of evolution. Some of these even have produced creation museums, such as Ken Ham in his "Answers in Genesis" series and Henry Morris, founder of the "Institute for Creation Research." Any student of creation may go to the above mentioned websites and obtain more literature than most could ever read. These creation researchers have sound, solid evidence which supports the conservative biblical creation account for a young earth. In fact, Dr. Henry Morris stated before his death that he knew of thousands of credible scientists, fully credentialed with post graduate degrees from accredited universities, who have become convinced Believers in recent creation or a young earth **(2)**.

It is not hard to understand that the science world wants to use time as their ally for understanding the unexplainable creation of a Divine God. When the Lord is omitted from the creation formula, then there has to be another story for people to believe. When a person makes up a story to get people to believe, they must continue to change the story as new conflicting evidence emerges, in order to keep the skeptics in their camp. That is exactly what has happened with the theory of evolution. As new scientific thoughts surface, the evolutionists must continue to change the story to make it more believable and credible. Adolf Hitler often stated that "if you tell the same lie often enough, then most people will believe it."

Some of these so-called experts of the scientific world continue to speak on the age of the planetary structure of being billions of years old, whereas, their evidence does not support those conclusions. Please read the evidence of what some accomplished scientists are declaring about the age of the earth.

For example, Saturn's beautiful rings look bright and clean. Billions of years of space dust should have tarnished the ring's eye crystal into a dull fray long ago. Saturn's moon Enceladus continues to jet ice from 101 geysers near its south pole. Secular astronomers face a challenge in explaining why all this material didn't exit Enceladus millions of years ago and why this little moon still has enough energy to fuel its many geysers.

Next, some planets, such as the moon emit too much heat to fit the billions of years model. Neptune emits twice the heat it receives from the sun, and Enceladus emits some ten times the heat secular scientists predicted. If God made Neptune and Enceladus relatively warm in the beginning, they should still have extra heat after only thousands of years, but not billions **(3).**

There seems to be plenty of good evidence that some progressive evolutionists are still seeking to use same old tactics of deceiving people, in order to perpetuate the deception that so many continue to publish. These deceptive ploys continue to undermine the scientific truths that have been discovered by several conservative scholars. The classroom textbooks are having a profound impact upon what the culture actually believe regarding the age of the earth. God's people need to read some of the conservative journals that being published today.

Another way to arrive at an understanding of the truth about the earth's age is to consider the facts from several trustworthy sources. These sources will include the Bible, geology, biology, chemistry, and historical data. Several quotes of current conservative writers are included in this biography that appears in the most recent journals and publications.

A. Geological Factors

After much serious study on the topic of evolution, even Charles Darwin began to have many suspicions about the theory that he had adopted. Darwin said, "As by this theory of evolution innumerable transitional forms must have existed, why do we not find them embedded in countless numbers in the crust of the earth? The number of intermediate links between all living and extinct species must have been inconceivably great!" **(4)** Fossils of the transitional animals simply do not exist anywhere in our world to support an evolutionary chain. If any such creatures did exist, then they would be put on display in a museum somewhere. The best that can be found from any period are scientific mutations. The absenteeism of even fossils of these intermediate animals disproves the theory of evolution.

Another geological fact that supports a Divine creation is the fact of soil and rock erosion that has taken place over the years. Modern agricultural practices in the twentieth century have somewhat slowed down the process of erosion, but the rate has been tremendous during the last one thousand years. The erosion rate of the earth today would have most all the soil and all the fossil layers of rock washed into the seas in less than 15 million years, even if one never considered a huge worldwide flood. When the erosion from such a fact of history is calculated, it would likely cut that estimation seriously. A worldwide flood lasting thirteen months with the subsequent erosion, sedimentation, twisting and contorting of the earth's crust, and rapid burial of plants and animals is a much more scientific explanation of the geologic data (coal seams, fossil graveyards, canyons, etc.) than the uniformitarian myth **(5).**

The moon is receding a few inches each year. That means that the moon would have been so close to the earth a few million years ago that the tides of the ocean would have destroyed the earth twice each day. The moon draws the tides of the seas. This geological factor would certainly rule out the million and billions of years for the age of the earth. Another significant factor concerning the moon is the amount of dust on the moon's surface. The small amount of dust would indicate a time frame of less than ten thousand years since its formation.

The millions of years were factored into the initial landing for Americans on the moon on July 20, 1969. Several American scientists believed that the moon was covered with hundreds of feet of moon dust. They had the manufactures of the space craft to add long adjustable legs onto the spaceship, which would emerge just prior to the landing to penetrate down through the dust that would keep it above the dust. However, to their surprise there were only inches, instead of feet of dust on the surface of the moon. This evidence speaks to the point that the moon is only a few thousand years old and not millions nor billions of years old.

One popular question that most often arises over the age of the earth is that of a worldwide flood. Did God submerge the entire earth under water, or did He use all the water to flood a portion of the known world? Some have suggested the impossibility of a world-wide flood of forty days, since water did not exist to rain for such a period of time. Even if, such water were available to rain that long, was there enough water to cover the highest mountains which are four to five miles high? It would be safe for any Bible student to believe that the biblical record is correct and look for understanding that would explain such a phenomenon.

Prior to the flood, it had never rained from heaven above according to **Genesis 2:5–6**, but a mist supplied all needed water for the life of man, the animals and vegetation. Consider the weight and the pressure of water and what such a volume could have done to the earth's surface. With the stress of hundreds of feet of water falling over the continuous period of forty days and nights, water can create what is referred to as a continental tilt and uplifts in geological structures. That means that as the waters begin to recede with most of the weight being borne upon the lowlands that the pressure caused an increase in the mountains, thus causing the waters to run faster and faster from the surface to the lower levels, which formed the seas and ocean basins. Actually, the valleys sank lower and the mountains rose even higher following the flood from the weight of the water.

Erosion was so intense from such rapid water runoff, which helped to further reshape the entire topography of the landscape. This uplifting effect would further help to explain the fossils of animals at such high altitudes. There were probably very few similarities between the landscape after the flood than that of Noah before the flood. All things had changed. For example, the Grand Canyon,

Niagara Falls, Victoria Falls, and other such mammoth landscape wonders could have easily been formed during the after flood runoffs.

Then where did all the water come from that created such a massive size flood. Scientists have suggested that a water canopy existed above and below the earth that completely enveloped the earth and its inhabitants. This would be consistent with the Bible account in **Genesis 1:7**, which states that there were waters above the earth and below the earth. This layer of water surrounding the earth would serve as a protective barrier against the harmful radiation of the sun, while keeping the temperatures at a very pleasant level. This produced a greenhouse effect with a climate controlled environment with ideal temperatures, an oxygen-rich setting, proper amounts of water, where plants and animal life could thrive. A pleasant mist watered the earth's surface each evening or morning. God had provided a perfect environment that had built in protection for man, the animals and the plants.

Dr. Joseph Dillow has calculated how much water barrier it would be physically possible to suspend above the atmosphere as a blanket around the earth **(6).** He has suggested that the barrier which consisted of about forty feet of water vapor, was sufficient for the flood waters for forty days and nights. According to Scripture, the windows of heaven **(Gen. 7:11)** were opened, which indicated the hand of God released the waters through openings allowing it to literally pour out the water canopy around the earth and upon the earth. That water canopy does not exist today. According to Scripture, it did not rain upon the earth until the Great Flood in Noah's time. **(Gen. 2:5).**

Not only did it flood from above during the forty days, but the Bible says that the fountains of the great deep were broken up **(Gen. 7:11).** Apparently, large openings in the earth's surface opened and water began to gush out. Some suggest that the earth contained an inner supply of water that was under pressure, maybe in a steam form, which broke open at God's command like some type of volcano. The evidence of this is explained in the interspersed fossil layers in the rock record-layers that were obviously deposited during Noah's flood-then it is quite appropriate to suggest that these fountains of the great deep may well have involved a series of volcanic eruptions with prodigious amounts of water bursting up through the ground **(7).** It is possible that much of this water was trapped inside the earth on the third day of creation, when the Lord rearranged the universe with dry land and waters. Tons and tons of water began to rush violently upon the earth from every direction. As these massive volcanic like ruptures occurred in the earth's surface, the earth began to take on a new shape even during the early days of the flood.

During and following the flood, the new topography took shape from the weight of the water upon the earth's surface forcing the mountains to a higher level. This would become crucial to change the wind currents, in order, to bring about the rains following the flood. In today's world, these major winds and high mountain ranges are a very important part of the cycle which brings rain on to the continents **(8).** There are compelling pieces of scientific evidence today that support the canopy **(9)** water vapor theory, the radical shift in the topography, the great world-wide flood and the massive erosion following the flood water runoff.

Another supportive piece of evidence for the universal Flood are the fossil records and the layers of sediments found around the globe. There are places all over the earth where fossils of animals have been discovered in the same strata of rock, which indicates that these animals died at the same time. This would support a catastrophic event bringing wide destruction to large populations of animals during the same period of time. Fossils are formed when the skeletons of animals are buried under water for long periods of time.

Consider the fact that many fish are found fossilized in the same levels of rock formation. It is widely known fact that when fish die that they generally float to the surface of lakes, streams, rivers, and oceans, where they are consumed by scavenger animals. They rarely sink to the bottom and become fossils. Throughout the world there are thousands upon thousands of fish fossils that suggest that they were buried in a catastrophic event such as a flood. The multitude of fish fossils in the same layers of rock would not be possible without a major happening such as a worldwide flood such as the flood in the Bible in the day of Noah. Some scientists speculate that the fish fossils were frozen for millions of years; however, nature teaches us that the earth goes through periods of thawing throughout most of the world. This means that the fossils could not have been preserved for million of years.

Additionally, the universal flood is supported by the disappearance of the large animals often referred to as dinosaurs. It is highly unlikely that all of these animal species would have disappeared at the same time apart from some major event that destroyed them. Geological evidence and fossils consistently conclude that most of these mammoth type animals became extinct about the same time, rather than dying off slowly over millions of years as some secular scientists have concluded. The deluge of the great flood explains the sudden end to this large group of animals from the earth. Their fossils have been found buried deep in the earth surface indicating again some type of catastrophic event such as the flood.

Another more recent phenomenal happening that supports the young earth belief was the Mt. St. Helen's volcanic eruption in 1980. The molten rock created a steam blast with the force of 20 million tons of TNT. Within a matter of a few days, the avalanche had deposited up to six hundred feet of layers of new rock and debris-layers like those which secular scientists claim to be billions of years old **(10).** This geological formation in the state of Washington proves that rocks and rock layers may not be as old as some scientists have been reporting. This single event of rapid volcanic activity proves that it does not require millions of years to produce several inches of volcanic ash.

In fact in ***The Acts & Facts Magazine*** January 2017 **(11),** the magazine reported on a story that the evolutionists have printed regarding the age of the Hawaiian Island chain being billions of years old carries many problems that exist in their report. Those attempting to prove that these islands were formed millions of years ago when the Pacific plates moved across a hot spot of volcanic activity would have caused a development at a rate of a few inches per year. As these volcanoes became active they left a trail of progressively older volcanic islands **(12).** However the young earth believers, such as Geophysicist, John Baumgardner, demonstrated that the plates would have moved much more quickly during the Flood, at rates of several yards per second, creating the Hawaiian Islands only a few thousands of years ago **(13)**.

Baumgardner goes forth to explain landforms of these islands have lava tubes and waterfalls common on all the islands, which suggest all are much younger. Again the Geophysicist declares that the lava tubes transport molten lava, but today are merely hollow cave pipes. These hollow pipes cannot exist for millions of years before collapsing **(14)** All the islands have lava tubes, steep valleys, steps with dramatic waterfalls without many years of erosion. This indicates that these islands are very young when compared with current rates of erosion. The erosion process would completely destroy the islands in only a few hundred thousand years. The erosion rate reported is seventy-six miles of erosion in one million years, would completely eliminate the islands **(15).** Those proponents of the billions of years old earth theory have been proven wrong in practically every one of their arguments about the age of the earth, yet they continue to cling to their fabricated stories, especially in the textbooks in public schools. This is their only answer to deny the universal flood in the days of Noah by adding the millions of years theory.

Thus, geological evidence does not contradict the biblical account of creation and the great Flood. The Bible is far more consistent with fossil records and other geological formations around the world than the view of the theory of billions of years of slow development of the earth and the animal kingdom. Basically, the science of billions of years is in conflict with the Bible accounts of creation and a worldwide flood.

B. Biological Factors

If the earth were billions and billions of years old, then there would be trillions and zillions of people upon the earth today rather than a few billion. The population of today's world certainly supports the biblical account of creation. Around two thousand years BC the earth was destroyed by a worldwide flood leaving only eight people. Over these past four thousand years, it is logically true that the population is less than six billion people instead of trillions and trillions had man lived for millions of years as some scientists claim.

A popular biological question is "What about dinosaurs?" Did they really exist? It was not until 1841 that Sir Richard Owen, a paleontologist, coined the word dinosaur, which refers to giant size reptile creatures, which lived upon the earth. Some scientists suggested that these large bone animals lived long ago and could have been thirty or forty feet long with long necks and tails. Modern science says that these creatures roamed the earth millions of years ago before man. They contend that these prehistoric creatures mysteriously died out for some unknown reason such as an ice age.

Do the creation events of Scripture shed any light upon the biological development for such large animals like dinosaurs. Actually, the Bible teaches that the large animals were created on day 6 just prior to man, but it does not give any details to the size of the beasts that roamed the earth. It does seem impractical to believe that such large size animals as suggested by science could have inhabited the ark during the flood unless they were small versions that God sent into the ark. Those that were not on board the ark and were not aquatic types like fish, likely drown during the flood and some of their remains were preserved during the flood sediment period.

In the discussion on geology that proceeds, the great flood and its after effects made tremendous changes such as massive soil erosion, uplifts of mountains to high altitudes of freezing temperatures which preserved many bones that verify some large size animals once inhabited the earth. Since many of these findings were preserved in cold storage, it has made it impossible to determine the actual dating. The question of when such creatures actually lived upon the earth must be measured in light of the biblical record of time. When a person determines the Bible is true and accurate, without any mixture of error, then the answer to the life of dinosaurs on earth will become more clear.

In the book of Job, which many scholars believe is the oldest Bible writing, there are several references to large animals that God created and with each account, the writer describes these at the same time that man lived. For example in **Job 40:15–18**, God asks Job about one such animal that resembles a dinosaur, which is called Behemoth. This creature ate grass like an ox, had powerful muscles, a huge tail like a tree, bones like tubes of bronze, and limbs like iron. The mammoth creature was fearless of all other life and works of creation **(Job 40:23).** It is obvious that God asked Job about an animal that he was familiar with in his day.

An even more enormous size creature with fascinating features was called Leviathan **(Job 41:1–9),** which was a sea creature that shot fire from his mouth and blew smoke from his nostrils. This creature had an armor of scales that completely covered his body. Leviathan could have survived the flood without being inside the ark, since he lived in the water. He is also mentioned twice in the **Psalms (74:14 and 104:25–26)** and also was called the twisted serpent by Isaiah **(27:1).** It would be safe to say that man has then lived alongside of some very large sized creatures. In fact, there have been several fossils, teeth and large bones uncovered in recent years by archeologists that places the footprint of man and dinosaurs in the very same strata of rocks, which further indicates that they lived at the same time. Yet many scientists continue to insist that dinosaurs lived billions of years before man came into existence.

When one understands and accepts God's creation account, then he must agree that man and dinosaurs did indeed live together from the earliest of creation. The earth was created with a perfect environment, where men, animals and vegetation grew rapidly and likely far exceeded the present day size. With an abundance of water, plenty of clean air, proper light, without diseases, no pollutions, the perfect balanced earth produced large size plants and animal life. The dinosaurs and other large animals were needed to help keep the vegetation in balance, especially while the human population was small.

Since animals were not consumed for food until sometime after the flood, then they reproduced and thrived well with ancient mankind. God gave man the opportunity to eat the meat in **Genesis 9:3**, following the flood. If all of the biblical record can be trusted, then likely the dinosaur like creatures perished in the flood or within a few hundred years following. A small size pair could have been preserved on the Ark and that generation could have died out later or mutated to a smaller version later. Some scientists today are claiming the smaller mutations are related to the lizard family.

These speculations about dinosaurs could easily fit into the creation plan of six thousand years. Many modern secular scientists refuse to accept a universal flood that covered all the earth, with huge sediments and shifts of the earth, thus rejecting the biblical record. They add years upon years to explain the works of nature over a period of one year instead of the great flood.

Most modern science theories about the age of the universe utilizes a carbon dating system called carbon-14 to determine the age of animals and man from discovered bones and fossils. Carbon-14 or C-14 is radioactive and manufactured in the upper atmosphere from sun rays. Almost everything on earth contains carbon as it is recycled from plants to animal life. C-14 is 14 times as heavy as hydrogen, but falls apart very easily into hydrogen, especially when a plant or animal dies. So the scientist measures how long the plant or animal lived by the amount of carbon left.

The problem with carbon dating exists when the scientist assumes that he knows how much carbon was present before death in order to determine the actual time since death. The assumption is generally made that the organism had the same amount of carbon years ago as a similar one does today. This is a false assumption, since there is much more carbon present today. The only change in the scientific consideration for changes in amount of carbon are allowances for the industrial revolution with its huge masses of burning coal. The more one has of a radioactive substance, the more there is to decay—that is, as more enters a system, the rate of leaving the system increases **(16)**.

Carbon dating can be a good test when it is used properly; however, when it is used to support a preconceived theory, the test is simply not scientific. Those that use carbon dating to disprove biblical facts have predetermined the results to fit their theory. Even those modern scientists that have discovered many discrepancies in the carbon dating system, seem to be satisfied with letting the errors stand. Even with all the fallacies in this carbon dating system, many scientists are unwilling to deny the problems, since they have no other answers.

Even evolutionary scientists acknowledge that radiocarbon dating cannot prove ages of millions or billions of years. Dr. Vernon Cupps, PhD in nuclear physics from Indiana University stated that radiocarbon is an unstable form of carbon that spontaneously decays into nitrogen over time **(17).** These experiments have simply failed to match the smell test, but are still found in the classroom textbooks as true.

These discoveries have determined that the rate of inflow into animals is not constant, just like the rate of outflow is not constant. In fact, some measured rates on trees have revealed that there are many more inconsistencies of the rate of carbon build up in a plant than ever thought. It is a very complex system that holds many problems that have not been resolved by science. The former atmosphere certainly had a slower rate of carbon build up than the present earth. In general, this would give a slower rate of breakdown of carbon deposits, which would significantly reduce the length of years of the universe.

Another variable that modern science has ignored is the water vapor canopy period that protected the inhabitants and plant life from these radioactive materials. This pre-flood period would definitely destroy the current methods of carbon dating. The carbon dating system is much too inconsistent and unreliable to be trusted, but it happens to be the best that the scientific world has to offer at this stage to reject the biblical account of a young earth.

C. Historical Factors

The dating of the modern calendar should provide proof that all of history is linked to Christ and His redemptive work on earth. The calendar dates back to this significant event of the cross, which makes a strong declaration to the world. If man can trust these dates and their link to the Bible facts, then it should not be any more difficult to understand that the rest of Bible history is also accurate and trustworthy. From all known accounts, the ancient world had no problem relying upon the biblical accounts of dating the world's birth at 4000 BC. There was plenty of intelligence among mankind in past generations that contended for the calendar dating making the Coming of Jesus Christ as the most important date in history when He died for humanity. They saw this as the most central date in the life of human beings, so the death and Resurrection of Jesus should be the biggest event in human life. Thus, they dated things before Jesus as BC (Before Christ) with all things After the Death of Jesus as the AD (After Death) period.

In biblical history, there are ten generations from Adam through Noah and the great flood, which covers a period of 1,656 years according to Genesis 4. There are 1,900 years from the creation of Adam to Abraham and 2,100 years from Abraham to Jesus, which totals to approximately 4,000 years. Many contend that the years of Genesis are different periods of time than the modern-day years. Those usually insist that man could not have lived nine hundred years before the flood. The time and seasons are set into motion in day 4 of creation and there is no biblical record of any change. Due to the fact that God's original creation in Genesis was a perfect environment without flaws, imbalances, diseases, pollutants and many of the present-day destructive evils, it was entirely possible to live for nine hundred years or even longer.

A serious conflict in the dating system of evolutionist lies between their claim of the earth to be billions of years old and the scare tactic of overcrowding of the earth. Proponents say quite the opposite on these issues. If the earth is billions of years old, then why is the world population only 7.5 billion people. Science evangelist, Dr. Kent Hovind studied the earth's population and discovered that there are approximately seven acres per person. He contends there are approximately 37 billion acres of land on earth and about 5.4 billion people when his study was performed. Hovind further contends that the entire world population could fit into a circle with an eleven-mile radius such as Jacksonville, Florida **(18).**

As one begins to study the history of civilization and the population of the earth that it becomes increasingly clear that the biblical record is accurate and sensible. Every other system or theory has serious problems and conflicts with itself.

The oldest known historical records of man are less than six thousand years old **(19).** Are there not writings before this period, if man had lived for even millions of years upon the earth? Early man was always clever enough that surely he left messages scribbled down to share with his fellow man. Strangely enough, no messages of any sort of a date beyond six thousand years seems to exist. This fact alone makes a powerful statement about the age of the earth and mankind.

D. Ecological Factors
(The Universes Demonstrate a Balance)

Everything in nature seems to work in concert with other things that exist in nature. For example, the planets all have their place in an organized manner that compliments each other and continue to rotate in their own particular patterns. One does not observe any great change as the planets in our solar system keep moving around the sun in a very exact pattern. This would suggest to any serious student of science that something is behind this exact organization of matter.

Indeed the Bible also states that everything was purposefully placed in this world by the voice commands of the Lord God. In **Genesis 8:22**, the Bible declares that it is God who is governing the universe as it says, *"While the earth remains, seedtime and harvest, cold and heat, winter and summer, and day and night shall not cease."* This statement occurs after the Great Flood as a promise to Noah and his family. Since the Lord God does not go back on any of His promises, man can trust that promise is still true. Then it is God who keeps order and organization to this entire universe and to all the universes that He has created. None of this just happened without a Supreme Designer. In the teachings of science, everything has an origin and the Bible simply states that God is the author of all things. In **Colossians 1:16**, the Scripture declares *"For by Him were all things created, that are in heaven, and that are in earth, visible and invisible,* *whether they be thrones, or dominions, or principalities or powers: all things were created by Him and for Him."*

There are laws of science that clearly demonstrate that the universe is in perfectly balanced with itself. For example, there exist a well-documented scientific balance of the exchange between plants and animal life. This process of exchange of photosynthesis in the plant life has to have carbon dioxide in order to produce its food for growth. During this complex process of producing plant food, the plants use the carbon dioxide and manufactures oxygen, which is absolutely essential for human life. Could it be possible that this unique and complex process just evolved into existence or was there a Genius behind this magnificent development?

Is there a scientific answer to this complex process of photosynthesis and millions of other scientific facts that have been tested over thousands of years? It certainly points to a definite planned world with an awesome Architect. Once again, the Bible provides that answer in Colossians, as it informs the reader that Jesus Christ is that designer of the worlds. The Apostle Paul testifies it was Jesus who did it all. *"For by Him all things were created that are in heaven and that are on earth, visible and invisible..."* **(Col. 1:16–17).** The statements from the Word of God are very emphatic that nothing accidently just happened upon its own power, since the Bible gives credit to the voice commands of the Lord God forming each item that came into existence. Over and over in the Psalms, Isaiah, and elsewhere in the Old Testament books, the Bible asserts that God called forth everything physically from nothing. Everything that God created had a relationship with other things that He created such as animal life with the plant life being so intricately connected with carbon dioxide and oxygen. These acts of creation are simply too complex to have happened haphazardly. How would the animal kingdom have known that it would need plant life for food and its very existence with oxygen?

There is an observed balance in the insect and animal world that keeps things in check. Also, every species of animals and insects have predators that keep their population in balance. It does not take a Genius to notice that things in nature are generally in balance with all others things of the created order. How could order and organization come from a world developed from chaos or chance, which hold the non-creationist view of life on earth?

Today there are people that are greatly concerned about man-made conflicts with the natural environment. In the early part of 2010, a huge oil rig accident in the Gulf of Mexico sent many people in that region of the world into a panic mode as crude oil begin to corrupt the beautiful waters of the Gulf of Mexico. The owners of the oil operation, British Petroleum or BP were reprimanded for their careless operations that led to the explosion of the rig. They were fined $20 billion by the US government and paid untold other billions to people that depended upon the gulf for a livelihood.

Environmentalists and millions of others quickly condemned the BP operations. Many stated that the accident had completely destroyed the fishing business in South Louisiana and declared it would take a decade or more for any kind of recovery. Within a period of three to four months after the drill hole was plugged, most of the oil disappeared from the surface. Within one year from the explosion things had really cleared up in the gulf. There is no way that this writer can condone the careless actions of the oil company; however somehow nature dealt a helping hand to all the citizens that depend upon the gulf like fishermen, tourists, shippers and others. There is no doubt that some of the pollution from this accident could remain in the ocean floor for many years to come.

It is certain that the churning of the ocean waters that are pulled by the tide of the moon had an enormous impact on the cleaning of the waters of the Gulf of Mexico. This same type of ocean agitation and churning assist the environment in cleaning the waters of it dirt, grime, oil, salts, and other substances that continually get into the waters. This ocean churning has always been helpful in cleansing the ocean waters. Now, the salts from the ocean have been determined to be healthier for human consumption than the traditional salts mined from land regions. The churning of the ocean is something like the home washing machine action that removes dirt and grime from soiled clothing with constant agitation inside of the washing machine.

God placed these Laws of Nature or Laws of God into place whenever He created the worlds and all that exist. This demonstrates to mankind that there is a Mastermind behind the creation and the maintenance of the universes which Paul wrote about in **Colossians 1:16–17**.

If God has indeed placed balance into His universe, then that would extend to the more modern topic about "Global Warming" or sometimes called "Climate Change." Millions or perhaps, billions of people are confused on the status of Global Warming. This writer, like many others, have been researching this topic for over two decades but the facts and figures just don't add up. After some serious admissions in 2009, that some people had been intentionally adjusting the facts on this subject, there are far more skeptics today with the reports.

Thanks to these folks who admitted to tampering with the reports, in order to make climate change appear to be a grave problem. As a result of these admissions, there seems to be more people that have been checking out the facts. Recently, an *Associated Press* article by Arthur Max appeared in local papers on December 1, 2011, declaring that the US was foot dragging on the key issues concerning "climate change." Mr. Max further added, "For us in the developing world, the biggest threat, the biggest enemy, is climate change" **(20).**

The international annual Global Warming Conference with some fifteen thousand people in attendance was held in Durban, South Africa during the first week of December in 2011. Each year a group of so-called scientists and science activists gather in some country to promote the dangers of global warming by insisting that the planet is continuing to heat up. There is much scientific data that actually suggest quite the opposite. Those that compile temperature records during the last century suggest that there is no increase in temperatures, but there are definite temperature cycles that fluctuate from hotter to cooler temps in many places around the world.

It is interesting to note that these supporters of the fact that man is responsible for a continual rise in the temperature of the earth for years and then suddenly changed the emphasis to "climate change" from global warning. Here is the reason that the subject matter was changed. The truth is that the earth has gone through a cooling for the past several years and has actually cooled by 0.7 degrees in the past century **(21).**

There is another twist in the long reported problem promoting "global climate change" or "global warming" and this comes from the problem brought to bear with volcanic activity. It is widely reported and estimated that there are an average of around two hundred active volcanoes on the planet spewing out these gases at any given time every day. Consider, Mt. Pinatubo in the Philippines that erupted in 1991 and continued without ceasing for more than two years sending greenhouses gases throughout the earth. This single volcano spewed out more greenhouse gases into the atmosphere than the entire human race had omitted in the thousands of years that the earth has existed. The ash and gases reached around the entire planet every single day. This author gathered up some of the ash during those days just a few miles from the actual volcano in Angeles City on Luzon Island. The ash literally filled up rivers and streams, while burying homes under hundreds of feet in the immediate proximity of the volcano.

The global warming enthusiasts have been promoting international treaties among the nations of the world that would fund such a program. One suggestion made in 2005 proposed sending up satellites that would be equipped with huge fan blades that would help to cool off the planet. Most of the so-called solutions in response to the unproven theory of global warming are far out for most developed countries to consider supporting. Billions of dollars are being spent to prevent the escalation of global warming while people are starving to death in third world countries. Likely most of these dollars are coming from the US government. As of today no funding has come from China and other developing nations to operate programs to halt global warming. If these unproven ideas were true, then all the nations of the industrial world would have to become evolved in the effort to thwart such an escalation in temperatures.

Perhaps, the best known proponent for global warming is the former Vice President of the United States, Al Gore. This outspoken advocate for global warming appears at most of these conferences, suggesting funding for his global warming theory, while asking every nation to pass tougher standards for global emissions worldwide to slow down threats of the universe overheating **(22).** Gore was awarded one half of the $1.6 million for his selection prize at Bali, India in December 2007. Gore stated that the United States and China are the biggest producers of greenhouse gases. He concluded by stating that these two nations should take the lead in solving this problem. If there is no global warming, why should anyone attempt to solve anything that does not exist. It remains unclear that any facts exist to support Mr. Gore's theory.

Shortly before President Barack Obama left office in 2017, he stated that the biggest problem facing the world was the matter of Climate Change. People instantly began to inquire as to where he was obtaining any evidence for such an outrageous comment. Shortly after Donald Trump became President of the United States, he withdrew any US financial support with the international community on the idea of Climate Change issue in the Paris Peace Accord, because America had been paying 90 percent of the research on the issue. One of the candidates running for President of the US, Senator Bernie Sanders said in August, 2019 that the most primary issue facing America is not economics nor illegal immigration, but is Climate Change. This illustrates how misinformed people are about this matter, since there is absolutely no scientific facts to support this unproven theory that Climate Change exists.

In September of 2019, a young 16 year old girl from Sweden suddenly appeared on the world stage with a well written script declaring that the earth was currently in a major crisis. Perhaps, she was being used by the Global Climate Change alarmists to produce their message that everyone must immediately convert to Green Energy to save the earth. Her message created instant fear in the hearts of so many people that all folks that disapproved of her message were labelled either reckless or misinformed about the immediate crisis.

The young teen was so convincing with her message that the earth would likely have a meltdown within eight years. Many public schools across America allowed their students to miss classes while holding demonstrations in the streets of major cities to convince others of this urgent world crisis. The young Swede speech to the nations of the world charged world leaders to act immediately to save the planet, since this was presently a life and death crisis for the earth's existence. This would mean that all internal combustions engines using fossil fuels must immediately be outlawed. Additionally, it meant that gasoline powered cars, trucks, airplanes and heavy equipment would have to be replaced within 2 to 5 years with green energy fuels to resolve this crisis.

There are multitudes of problems of following these proponents with their unscientific theories. Not the least of these could be the prohibitive costs of trillions dollars of new tax money per year in America. Some economists suggested the cost for taxpayers would double or even triple above the current level. The irony of the story is that there is not a shred of scientific proof that Climate Change even exists. In fact, the last half century facts reveal that the earth is now in a cooling pattern of .7 of a degree, while these alarmists are declaring a current crisis has enveloped the entire planet.

Since the early days of the 20th century, perhaps even long before, there have been alarmists crying out for America tax dollars in order to conduct a remake of America for their own personal gain. The truth is that man cannot control the climate, since our Lord is in full control. Certainly, man can mismanage the earth and dirty it all up, but man cannot change the temperature settings nor make it rain when he desires. Since temperatures have been recorded, they have always gone through cycles of cooler to warmer temperatures.

Many "climate change" enthusiasts today are stating that there are stronger and more violent storms happening today actually prove that the earth is warming at a regular or faster pace than in the past. The scientific evidence does not prove this to be true, but rather a false assumption. In fact, true science proves that when there are two different air masses with different temperatures that the storms are most likely to become more violent and stronger. In other words, when a very warm mass of air conflicts with a very cold atmosphere of air, the storm is likely to be far more catastrophic.

Cal Beisner founder of the Cornwall Alliance for Stewardship of Creation states that more intense storms are produced during seasons when there are colder temperatures than normal. The warm movement of air pushing against the colder climate creates a far more violent storm **(24).**

So what does the Bible have to say or does it say anything about "global warming" or "climate change"? There are some biblical facts that ought to be considered by any serious student of the Word of God. Once again, most noteworthy would have to be **Genesis 8:22,** which says, *"While the earth remains, seedtime and harvest, cold and heat, winter and summer, and day and night shall not cease."* This is clearly stated that there will continue to be changing seasons, thus no need for alarm over climate change. This does not mean that anyone has the privilege of total disregard for the environment, but it does mean that some people may choose to ignore good conservation practices while others become alarmists. They intentionally alarm people when there is no danger. All people are responsible for their actions and the consequences of sin.

E. Theological Professional Positions

There are many trustworthy professionally trained scientists that are Believers in the Lord Jesus Christ, which should not be overlooked that have studied on this subject for many years that bring a needed perspective upon this subject. There are many such persons of the Christian faith that are qualified to speak on this subject that could easily be considered as experts on the age of the earth. Some of these that have labored for decades in serious study of this subject are those connected to several Christian Research groups, especially with those in America.

Ken Ham with the Answers in Genesis organization in Hebron, Kentucky, and his team of expert historians, scientists, theologians, astronomers, medical doctors, researchers, and a host of others that have a voice that deserves to be considered in this conversation. Ham concludes his remarks on the age of the earth by declaring, "Many scientists and theologians accept a straight forward reading of Scripture and agree that the earth is about 6,000 years old" **(25).**

Another creationist organization is the Institute for Creation Research with leader Henry Morris III, who has concluded that all of those who reject the biblical model "must believe" in the long ages for the universes, which is really a faith or a belief. Morris states, there is no other scenario that could suggest that the current processes that we now observe and upon which we now depend would have enough time to "evolve" from something less (simple) to that which exists **(26).**

Chapter 12

The Age of the Earth

1 Hovind, Ken, ***Dinosaurs Creation Evolution***, Creation Science Evangelism, Pensacola, Florida: 1990, pg. 5.
2 Morris, ***Acts & Facts Magazine***, July 2016, pg. 6
3 Thomas, Brian, ***Acts & Facts Magazine***, March 2017, pg. 20.
4 Hovind, ***Dinosaurs Creation Evolution***, pg. 39.
5 ***Ibid.,*** pg. 23.
6 Dillow, Joseph, ***The Waters Above***, Moody Press, Chicago, Ill.: 1981, pg. 120
7 The ***Answers Book***, pg. 119.
8 ***Ibid.***, pg. 122.
9 Snelling, Andrew**,** ***AFA Journal***, Is Genesis Really True?, February 2017, pg. 25
10 ***Acts & Facts Magazine***, Institute For Creation Science, January 2017, pg. 9.
11 ***Ibid.***
12 ***Ibid.***
13 ***Ibid***
14 ***Ibid***.
15 Ham, ***The Answer Book***, pg.68
16 ***Acts & Facts Magazine,*** Vernon Cupps, "Radiocarbon Dating Can't Prove an Old Earth" April 2017, pg. 9.
17 Hovind, Ken, ***The Answer Book***, pg. 36.
18 Morris, Henry, The Answer Book, pg. 160.
19 Max, Arthur, ***The Jackson Sun***, daily newspaper in Jackson, Tennessee, section A6, December 1, 2011.
20 ***Rutherford Plimer, Australian geologist professor emeritus of earth sciences at the University of Melbourne.***
21 Pachauri, Rajendra, UN Chairman of Panel on Climate Change in 2007, when Al Gore won the 2007, Nobel Peace Prize
22 ***Acts & Facts Magazine,*** Vernon Cupps, "Radiocarbon Dating Can't Prove an Old Earth" April 2017, pg. 9.

23 Beisner, Cal, ***One News Now***, Cornwall Alliance for Stewardship for Creation, daily report, June 2, 2013.

24 Ham, Ken, General Editor, ***The New Answers Book 1***, Answers in Genesis, Masters Books, Green Forrest, Arkansas: 2011, pg. 124.

25 Henry M. Morris III**, *The Book of Beginnings*,** Vol. 1, Creation, The Fall and The First Age, Institute of Creation Research, Dallas, Texas: 2012, pg. 34.

26 Al Gore, former vice president of the US, who was awarded the Nobel Peace Prize in 2007 with UN Rajendra Pachauri, UN chairman of Panel of Climate Change. E

CHAPTER 13

THE THEORIES OF CREATION EXPANDED

When one begins a study of a subject as broad as that of creation, there is such an overwhelming question of where does a person begin? Volumes of material have been written upon the subject from a scientific viewpoint, as well as, many volumes from the religious point of view. Today there seems to be a growing acceptance and toleration for the view of science over the biblical account. Admittedly, the Bible is not a science book, but it sure does include a lot of science. In some cases, the scientific view appears to clearly conflict with the scriptural version; however, in many other situations, there may be a simple explanation, which shows no conflict with the Bible.

Sometimes when the world pushes its beliefs before the spotlight, it causes God's people to rise up to the challenge to search for the truth and speak those truths that they have learned. This writer's intention is to call God's people to speak confidently the truth regarding the Word of God. It is important to be able to give a definite reply to those that are truly seeking for the truth or to help them to find the answers to their serious questions. Sometimes, that would include admitting not to have the answer, but helping them to find the answer.

If a student of the Bible is confused about the origin of life and how the worlds were founded, then that student will, no doubt, struggle even more with other issues in the Scripture. Once the Bible student has a degree of confidence about the foundation of life and how God carefully placed all things into His world, he will be far more prepared to find God's will for his own life. The writer of the book of Hebrews states that it is *"by faith"* **(Heb. 11:3)** that we can understand creation. He says, *"the worlds were framed by the word of God, so that the things which are seen were not made of things which are visible"* **(Heb. 11:3).** The case is clearly made that the Invisible made the visible by the Hebrew writer, as well as, other writers like Moses. This same writer concisely declares that the Word of God framed the worlds. Note that he specifically states the plural for *"the worlds"* **(11: 3)**. John the disciple of Jesus used similar terms in his Gospel of **John 1:1**, when he declared that "In the Beginning was the Word, and the Word was with God and the Word was God." In **John 1:14**, the writer declares that this proper noun (the Word) is the same for Jesus Christ from the Beginning of time.

The writer is not attempting to explain in detail all the varied beliefs that exist today, but simply to refer to those as they conflict with the biblical account of creation. However, for the fear of a misunderstanding of the discussion, a brief overview of some popular views will be presented in the introductory remarks and at other appropriate times. When one has a good sound understanding of the Bible, then he will be able to see the errors of the false teachings that the world has adopted.

When the banking institutions train their employees how to detect bogus money, they always train them how to identify the legitimate bills rather than teaching with phony money. They believe that if the employee can identify the real, then they will be able to spot the fake. This same principle holds true when a person learns the Word of God. The in-depth discussion will center upon the creation plan and seek to exalt and praise the Creator. The study seeks not only to inform the reader about documented facts of creation to increase knowledge, but to lead the reader to rejoice in his heart over the Divine God of creation. The Scriptures are to become a song to the heart of every Believer that spontaneously brings one to worship the Lord of Creation. These moments of truth discoveries are glory moments to worship the Creator.

No other single doctrine ties other doctrines together in a unity of consistency more than creation, since it touches upon all of the other doctrines of the faith with such a rippling effect **(1).** There is a definite network of doctrinal connections that helps the serious-minded Bible student better understand and appreciate the Bible account of creation. Doctrines such as the sanctity of human life, human freedom, laws of marriage, justice and many others have their foundation in the literal, historical understanding of Genesis **(2).** Perhaps, no other Old Testament scripture stands as tall as the Genesis creation story. There are far too many foundational truths that are vital to the whole of the Bible found in the creation account. When a student of the Bible desires to build a good foundation for truth, then other Bible truths will stand upon this sturdy foundation. There is no greater place to lay this foundation for truth than in the Genesis story. When man comes to understand the nature of God and His attributes, then he can more easily accept the facts presented. Thus this study will begin with a discussion of many of these doctrinal issues and beliefs that relate to creation and to the Creator.

For example, when a student comprehends the vastness of the universe, then he can begin to believe that the Creator is far greater than the creation that He so capably brought together. A Bible teacher once asked a group of pre-teens to write a brief statement on the subject of "How big is God?" One student responded by declaring that God was like an endless sea of tapioca pudding. While that might be viewed as a humorous or even a childlike reply, there is much truth in his statement. The God of creation is endless or limitless in His Being and abilities. Wow, what brilliant observations come through the mind of children!

When creation is used in this discussion, it will primarily be used as a reference to the beginning of the worlds and universes. Some scientists declare that there are trillions and trillions of universes that exist. That argument is not the topic of this discussion; however, if they do exist, then the God of this universe created those as well. If there is human life on other planets or in other universes, then Christ died on the cross for them. Since Christ died once for all sin, then man would be held accountable to getting the message of salvation to those that exist on other planets. Christ will not be crucified over and over again even for those in some remote part of our world, nor for any who might live in other universes. His wonderful work of grace is a completed work that has already been put on display for all to see and hear.

There are new discoveries of other planets every few years. In fact, on December 6, 2011, an article appeared in newspapers declaring that another planet had just been discovered that is eerily similar to earth in another solar system, which seems to be the ideal place for life. Astronomers say that it is situated in the Goldilocks zone, where it is not too hot nor too cold for life to exist. Scientists have declared that the temperature should be around 72 degrees, thus making it ideal for life to exist. The Goldilocks concept comes from the fairy story of the three little bears that came uninvited to the home of Goldilocks. Baby bear found that Goldilocks' bed, soup and chair were perfectly suited for him, while his parents found that things were unsuitable for them.

Without a doubt, one of the major issues regarding life on other planets would require a favorable temperature that is near to that of earth. However, almost all the planets that science has observed, there are only a few that qualify in their distance to the sun. Another favorable consideration is that water must be present to sustain life. If a favorable planet is out there and it happens to be in one of the other universes, it would likely take a few hundred years for man to build a spacecraft that would enable him to reach such a planet. Seemingly, the most favorable planet that most thought would have life was the red planet Mars. The United States has already successfully sent satellites to that planet, yet nothing was found to sustain life. Not even a trace of water has been discovered on Mars, the moon, nor any other planet beyond earth.

During the mid and even later 1900s, there were US citizens convinced that spaceships from Mars had already landed on earth loaded with little green men. While some were convinced that these strange visitors had already made contact with folks on earth, they were speculating more were coming to invade earth. All things are possible with the Lord, since He has made all things and all beings. If other beings do exist on other planets, then rest assured that the same God created them. God is the Maker of all things and all beings. As of now, no concrete evidence has been discovered of their existence. It is true that we have plenty of planets in all of the solar systems known to scientists today. Perhaps, God is simply showing off His handiwork to fascinate and to astound the more brilliant scientists. God may have created millions or trillions of different galaxies of planets to keep man intrigued until the Lord returns to bring this story to its conclusion or to assist man in discovering the Lord is his research.

The Bible says that *"the heavens declare the glory of God; and the firmament shows His handiwork" (Ps. 19:1)*. Thus, for the man who is searching and looking can find God in the midst of His work of creation. The work of the Divine God of glory certainly is not all a past tense work, but continues to grow every day. When a careful observer looks at a newborn baby, he must conclude that the Creator is still at work creating. Even though, the past creation is being analyzed in this study, God presently is creating and shall continue in the future to create and recreate many wonderful works. His work in redemption of sinful mankind is also a work of creation. The Apostle Paul said, *"Therefore, if any man is in Christ, he is a new creation: old things have passed away; behold, all things have become new"* **(2 Cor. 5:17).**

A. Ancient Philosophy (Mythology) Theories

Those of the ancient world held varied beliefs about creation, just as man does today. The most famous belief from the non-biblical world was that of the Babylonian people, or so-called Chaldeans, who believed in an ancient struggle of many gods for supremacy of the universes until one was able to subdue all others. After the great victory, this victorious ancient god established the present world from some of his own blood **(3).**

The ancients from the Greek empire explained the world's beginning from their Greek gods of love, chaos, Saturn and Tartarus with Saturn as the father of all their gods. Likewise, the ancient Egyptians had an explanation of how all things came into existence. One view from Egypt held that all life was created from gods and goddesses that had created water and all human life was formed from the germs borne in the water.

The philosophers from the ancient Phoenician world explained the cosmos hatched from a world-egg producing their world. Life moved from stage to stage of production. These were merely a few of those from ancient times that sought to satisfactorily explain the foundation of their existence substituting their gods and goddesses for the Lord God. However, when a student of creation begins to analyze today's beliefs of creation, he will likely find some very weird and similar views in the modern era. The modern explanation often includes a belief that there was a common ancestor from which all of life descended. Since there are millions of species of creatures present today, the species all must have developed by random chance over a period of millions or perhaps, billions of years that has brought human beings into this modern era.

In modern times, most of the explanations given from the secular world attempt to explain the origin of the world and life in terms that exclude any type of god. Today, there are several general approaches to man's belief of creation. The theories of the modern world—evolution, humanism, secularism—have one basic element in common: they deal in chance, luck, and randomness **(4).** For example, in 1995, the American National Association of Biology Teachers (NABT) went on record and made the following statement regarding the origin of life:

> The diversity of life on earth is the outcome of evolution: an unsupervised, impersonal, unpredictable and natural process of temporal descent with genetic modification that is affected by natural selection, chance, historical contingencies and changing environments **(5).**

The group of biology teachers carefully crafted a statement that claimed that the process of life came about without supervision (unsupervised) and continues in a blind and random pattern without any design or purpose. NABT concluded in 1995 that "there was/is no God of Creation." Thus, if man was randomly produced by certain genetic agents accidently coming together at a certain time, then man has no purpose and is going nowhere. Many people of today's culture actually believe that man is a product of blind chance, after certain things gradually aligned in a particular sequence.

Most Biology teachers are required to take dozens of science courses before becoming a teacher and have studied the complexity of life in scientific laboratories. This author obtained a BS degree in Science, while spending approximately a third of all his studies in science courses. It is amazing for a student of empirical or true science to become persuaded that all the molecules could line up by random chance to produce a frog, much less a far more complex man. The complexity of one human cell is often enough to convince most folks once they observe one under a high powered electronic microscope that man could not have evolved from a lower species of animal life. Someone long ago said blind chance producing a frog in a science laboratory would be like an explosion at a printing plant producing an unabridged dictionary. Wow, how could any intelligent being believe such an event ever happening by things lining up perfectly for such a happening by mere chance?

One worldly belief or philosophy is that God created all things, then wound it up tight like a big clock spring, walked off and left it winding down until some end point. In this particular view, God has purposefully disassociated Himself with all ongoing events, leaving the future up to the ingenious mind of man or to chance. A major objection to this view for the student who believes in God is the fact that the Bible teaches throughout that the Lord is actively engaged in sustaining His creation. This belief comes into serious conflict with the Scriptures, since God has integrated Himself into every aspect of His creation. He further has promised that He would not leave nor forsake those who call upon His great name through the repentance of sin. The place of prayer to God alone would deny such a false belief of the Divine Creator, much less the unlimited love of God that refused to allow man to remain in his sin without confronting him as He did after the very first sin in the garden. There are just too many serious problems and conflicts with this belief to make a credible argument for its validity.

B. Evolutionary Theory

Since this theory is far more advanced than any other theory on creation which opposes a Divine Creation, this study will devote more space in order to debunk some of the thousands of concepts that are attached to evolution. In other words, there is much in these teachings that do not match the other teachings of the same far-fetched belief. Public school textbooks are filled with these thoughts and suggestions in favor of this theory. This popular belief of the creation of man is that God had no part in creation, but rather a coincidental sequence of events lined up triggering a cosmic accident.

The modern science world today often describes a speck, which is referred to as a period somehow appeared out of nowhere and became energized into a spinning rotation. As the rotation continued and the energy level intensified, it blew up creating what has been referred to as the **"Big Bang Theory."** No scientist has ever been able to satisfactorily explain where the energy came from, since this theory conflicts with science, which declares that all energy must have a source for power. What gave the period its burst of energy? No one can say. Another non-scientific conflict that surrounds this particular view is the description of the particles spinning away from the explosion. Every scientist knows that all the particles that separated from the source would be spinning in the same direction as was the parent particle. However, all the planets in this universe do not spin in the same circular pattern, which means that the Big Bang Theory is totally erroneous and in direct conflict with the laws of physics.

The Big Bang theory relies on a growing number of hypothetical entities-things that have never been observed, such as inflation and dark energy, being the most prominent **(6).** This theory depends upon the idea of naturalism rather than a Supernatural agency that is found in the biblical account of Genesis. It all depends upon matter and things coming together at the right time and the right place. Again, these are ideas or theories that are substituted for the Supernatural without one shred of scientific evidence to support the hypothesis.

If there was a big explosion of sorts, then the laws of physics would lead us to conclude that the energy level that propelled the particles away from the parent particle would continuously be slowing down in its movement. However, there is absolutely no evidence that any of the planets are slowing down significantly, but rather are continuing to travel at approximately the very same rate and are still in the same course that they have always held. For example, the earth makes a complete turn on its axis every 24 hours and completely rotates around the sun every 365 1/4 days. There is zero evidence which would lead any scientist to conclude that there could have been a big bang that split the planets apart in the beginning. It is totally impossible to conceive that the planets created themselves by some unexplained explosion.

Perhaps, the biggest objection to evolution today is why did the process cease with no present evidence of evolution process? What changed this process after millions or billions of years causing the successful process to suddenly disappear? No answer can be found for this mysterious change. Another huge mystery centers upon absolutely no evidence of changes in the fossil records that are very plentiful all around the earth. Why have no fossil records of any intermediate creature of evolution never been found among the trillions of fossils around the globe? There are far more questions than there are answers abounding with the theory of evolution.

The Bible declares that it is He (God) who has made us and not we ourselves **(Ps. 100:3).** Just like God made man, He made the planets and hung them in outer space and it is the Lord that keeps them in their patterns today. "If there is no Creator, who sets the absolutes, why would any person still be following these Christian rules (about marriage, sex, truth, ethics, etc.)? Why doesn't man just do whatever he wants to do?" **(7).**

If no one is keeping things together, then things would eventually go flying off into many different directions. Nothing in our universe maintains itself, which is based upon the Second Law of Thermodynamics. This Law of Nature or Law of God declares that left alone, nothing cannot maintain nor repair itself. Over a period of time, things left to themselves will not improve, but rather will tend to break down. The idea that supports evolution is in contradiction to the Second Law of Thermodynamics and opposes almost every other Law that governs Nature. In fact, at the heart of the theory of evolution is one fabrication after another falsehood. Hitler along with most all dictator type governments in history generally removed all Bible teachings and begin to use the lies that evolution is built upon to teach their people.

When people are taught that they evolved from animals, it does not take long until man begins to act like animals. Morals quickly erode when people begin to believe that there is no God, which means the rules of life are no longer meaningful nor necessary. One final occurrence that happens is that human life has little or no value. Therefore, even murder becomes permissible in certain situations when a person is unable to make significant contributions to society.

When God is left out of the complex equation, there is simply no sensible answer and man's imagination begins to run wild. The thoughts of the Divine Creator are far superior to the mind of man as declared by **Isaiah (55:9).** Those that believe the spinning period theory which exploded conclude that this sequence of events happened billions of years ago, which is their suggested age of the earth. Now, even though millions of fossils have been found since Darwin's time, we still can't identify any transitional species-let alone "countless numbers" **(8).**

A closely connected leg that supports this theory on the origin of the planetary world is referred to as atheism for the beginning of life. This belief declares that human life and animal life started on earth without a Creator, but rather by many sequences of events lining up by random chance or by luck. Essentially, in this view of Time, it actually becomes the god for evolutionists. Followers of evolution believe that anything is possible given enough time. The view requires a lot of faith to believe this unproven theory.

Most of these followers believe that over a period of millions of years the animals advanced into quasi humanoids that lived in caves for many millions of years. The Bible does not teach that humans dwelt in caves for millions of years, but in paradise. Certainly these were times in the Bible where people would hide out in caves to avoid being found or captured, but these all were brief periods. God created the first man Adam with a brain and *"in the image of God"* Himself. The Lord placed His very character of righteousness into man, prior to the Fall of Man. From the beginning, man knew he had the desire of the Creator for his life with a heart to stay away from the tree of knowledge of good and evil in the midst of the Garden of Eden. The other tree in the midst of the Garden was the Tree of Eternal Life, which Adam knew was the accepted tree. The Creator had provided everything good for man in Paradise. Man had no reason to live like an animal, since he was created superior to the entire animal kingdom. Thus, man did not evolve from some inferior species of animal life, but was magnificently created by a loving God to enjoy the fellowship of the Divine Creator in the perfect garden of paradise.

The theory of evolution was hatched in the sinful heart of man that chose to deny the Creator access to his life. The theory came into popularity after the middle of the nineteenth century. Several people had advanced the view over a period of two to three hundred years, but none more noteworthy than a young British trained theologian named Charles Darwin, who wrote a book entitled ***On The Origin of the Species***, which was published in 1859. Darwin was highly trained at Cambridge with a degree in theology. As far as can be determined, this was the only earned degree for Darwin. He did study medicine at the University of Edinburg, but really became intrigued in the writings of his grandfather, Erasmus Darwin on the theory of evolution that was written the 1790s.

No single person did more to advance the cause of evolution than Charles Darwin with his writings, especially ***On the Origin of the Species***, during the 1800s. There are several writers today such as Stephen Jay Gould, Richard Dawkins and many others, who are continuously advancing this theory during the present time. Praise the Lord that at the same time these fabricated tales are arising from so many writers and speakers that the Gospel truth is also being passed along by many with a different viewpoint. Biochemist Michael Behe helps to deny evolution as he reveals the complexity of the human blood structure. In his book on Darwin's Black Box, Behe declares that "biochemical investigation has shown that blood clotting is a very complex, intricately woven system consisting of a score of interdependent protein parts." He further states, "The absence of, or significant defects in, any one of a number of the components causes the system to fail: blood does not clot at the proper time or the proper place" **(9).** Every function of the human body functions with other parts of the entire system as all work in tandem, which prove that blind chance is not a part of this equation. A Genius planned and manufactured the complex system and left nothing up to chance.

Spontaneous generation and Greek theology had already been taught for several centuries before Darwin began to develop a new twist to these old philosophies. He was heavily influenced from the ideas of his grandfather, who began to speak and write about his views on evolution. Actually, these teachings on evolution existed during the time of the ancient Greeks. For example, Aristotle (384–322 BC) believed in a complete gradation in nature accompanied by a perfecting principle from imperfect to the perfect with man at the top of the ascent **(10).** Many minds openly embraced this new theology or religion from the ancient world, which advocated that all of life evolved from a one cell organism in an upward spiral. Others who believed in evolution that impacted many others were people like Francis Bacon (1561–1626) and the first biologist was a French naturalist, George Louis Leclerc de Buffon (1707–1788) **(11).** In a much later time, Darwin's *Origin of the Species*, which was simply a printed theory that led to many other speculations in the later part of the nineteenth century and early twentieth century.

Darwin attempted to defend his theory of natural selection by stating the fittest would survive through the process of mutating or adapting to different environment, thus producing another species eventually. In the chapter, *Responding to the Objections* from those that opposed his thoughts, Darwin stated that they had not taken the trouble to understand the subject of natural selection **(12).** He further concluded that when the environment or physical conditions change, that those that survive will be forced to change in some manner, which will eventually lead to a new order. A critic insisted, with some degree of mathematical accuracy, that longevity is a great advantage to all species, so that he who believes in natural selection must arrange his genealogical tree so that all of his descendants have longer lives than their progenitors **(13).**

Darwin answers his critic by stating that it will occur on its own, without anyone aiding the process by nature or natural selection. It is true that many animals and organisms have experienced some form of mutation, but no dog has ever mutated into a cat type of creature or vice versa. By the end of that same chapter, Darwin is further defending his writings on embryonic transformations that can quickly cause huge changes in the appearance of animals, such as creatures that immediately go from no wings to wings, which he refers to as abrupt changes.

In the early part of the twentieth century, Darwin's writings were like a lightning rod that flashed around the world. About sixty-five years later, the same book set off another battle when a public high school teacher named John Scopes at Dayton, Tennessee began to teach the Origin of the Species with Darwin's theories in his public high school classroom in 1925. This resulted in one of the biggest legal cases in the nation's history up until that date. The citizens, who opposed the teachings of Scopes eventually won the case with popular attorney, William Jennings Bryant defeating the defense of attorney Clarence Darrow. Scopes was fired from Dayton High School and fined $100 for teaching falsehoods. He moved to Louisiana where he died a few years later, so the issues seemed to be put to bed.

Unfortunately, the belief was vigorously resurrected in 1959 by popular advocate and professing atheist Madeline Murray O'Hair. She formed an organization of atheists that begin to attack Christians, churches, Bible readings in public schools and prayers in the classrooms that led to the removal of these in many public school classrooms across the nation. The impact of O'Hair and other groups cost Christians many of their freedoms over the next few decades. O'Hair with some of her family suddenly disappeared in 1995, as well as, $630,000 allegedly stolen funds from the atheist organization. In February of 1998, new evidence linked the O'Hair group to a $608,000 money transfer from New Zealand and $100,000 worth of gold coins abandoned in San Antonio **(14).** It is interesting that only an estimated 4 percent of Americans still consider themselves atheists according to the Princeton Religion Research Center. How can such a small portion of the nation's population impose their beliefs and teachings upon the rest of society? This sounds like the tail may be wagging the dog. Perhaps, this has happened since so many churchgoers cling to a faith that does not match their lifestyles. These

Believers are no longer certain or firm about their beliefs of how that things were created, including God's highest order of mankind. It is sad that the present day culture is so uncertain about something that has remained certain for prior generations.

Even though the view was labeled a "theory" without any facts to support it, quickly the theory began to gain acceptance in many parts of the world. Again, Hitler said, "If you tell a lie often enough, then people will begin to believe it." So this old lie that declared man had evolved from a family of apes, who had evolved from a lower species of animal life has widely become accepted by the present society as the new science. So this old "theory" has now been accepted as a fact in most colleges and public educational systems around the world. In fact, the likely exceptions to the case are those conservative church members that hold to the inerrancy of the Scriptures. Unfortunately, there seems to be only a few conservatives speaking out compared to the opposing version. However, it contains a foundation based upon some very flimsy, faulty or farfetched ideas that cannot be supported by the facts of true scientific evidence. Even if an atheistic scientist found a big boat (Noah's ark) on top of Mount Ararat, he might deny that it was proof for a biblical Flood **(15).** Atheists or evolutionists will simply not accept Bible based facts. The finding of facts related to the Bible would often be referred to as coincidences.

Everyone believes in something, but most atheists believe there is no God. It is impossible to prove to people that God does exist, if they believe otherwise. Generally speaking, all people hold a faith in something or someone. The Bible is an accurate record of God's wonderful creation of the worlds with man as His crown jewel of all He called into existence.

It is absolutely stunning that Science does not support evolution, no matter how passionately the scientists and educators insist on it **(16).** As much as, the proponents of evolution want people to believe that science supports their theory, there is simply no empirical or true science in their theory. The fossils that have been discovered over the years of some intermediate apelike man have all been proven to be fabrications manufactured by men who desire to help evolution to succeed.

This unsubstantiated evolutionary theory of blind chance contains ludicrous and dangerous import of ideas. If there is no God, then human life is not sacred and it is without any real value. This means that human rights are ridiculous to consider. The same type of mentality existed in the early part of the twentieth century in Germany during the Hitler regime, whereby human life was greatly devalued. Such thinking is in direct contradiction of the Word of God, and the intelligent design of mankind. The Bible proves over and over again that this fabricated theory is without any merit.

Evolutionists believe that man and apes evolved from common ancestors about 50–70 million years ago. They primarily base their beliefs upon rocks, fossils, bones and teeth found in different strata of rock sediments. The belief or theory, which has no scientific facts for support contend that the single cell organisms evolved upward into more complex creatures by pure random chance. In the "thinking" of evolutionists, if enough monkeys typed for long enough, eventually one of them would type a perfect unabridged dictionary.

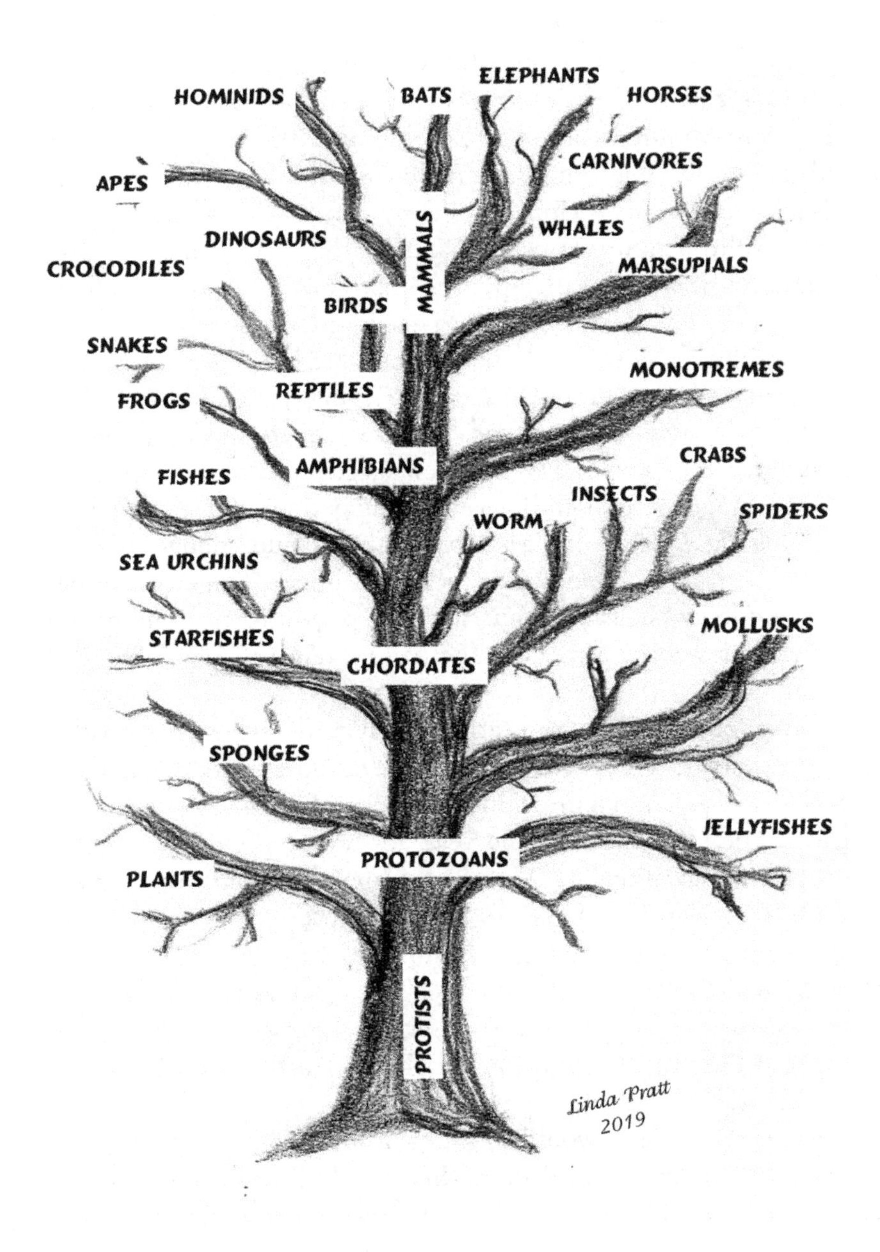
HOMINIDS
BATS
ELEPHANTS
HORSES
APES
CARNIVORES
MAMMALS
WHALES
DINOSAURS
CROCODILES
MARSUPIALS
BIRDS
SNAKES
MONOTREMES
REPTILES
FROGS
AMPHIBIANS
CRABS
FISHES
INSECTS
SPIDERS
WORM
SEA URCHINS
MOLLUSKS
STARFISHES
CHORDATES
SPONGES
JELLYFISHES
PROTOZOANS
PLANTS
PROTISTS
Linda Pratt
2019

To illustrate, consider the likelihood of just spelling the word "evolution" by randomly selecting nine letters from the alphabet in the proper sequence. The probability of success is only one chance in twenty-six by nine trials. This is equivalent to one chance in five trillion, four hundred and 29,503,679,000! **(18).** How in the world could any reasonable, rational, intelligent individual accept those kinds of odds for the development of any species? Furthermore, there would have to be a repeat performance of the same chances to produce another creature of the opposite sex to propagate the new form of animal life. These odds for this to occur are inconceivable for two such species, even over a period of billions and billions of years.

Modern-day evolutionist, such as Stephen Jay Gould and Niles Eldredge proposed a theory of "punctuated equilibrium" which declared that the chain of events of new species occurred in "spurts or gaps" of major genetic alterations after long periods of minor changes **(19).** This idea simply implies that nothing would change for long periods, then when nobody was looking a huge change to another species suddenly occurred. Thus the long period is the equilibrium and then followed by a rapid "punctuation." Again Time is the god of evolutionists.

Much of the billions of year theory gives the believers of evolution an explanation of why no man has ever observed any changing or evolving of an animal into another species of animal. For example, if it took one hundred thousand years for the process of a complete change from one animal to another, then the evolutionist can explain why no human has been able to make an observation which keeps their fabrication alive. This theory supports the **"macro evolution theory,"** which means a major change in a species into another species of creatures. Nobody has ever observed or proven a major change of a species from one form to another. True scientific discoveries simply are not there.

Many Christians would agree to minor changes in animals, which is called **"micro evolution theory."** This has been observed when a pair of black spotted dogs produce an offspring dog that does not look like its parents, but is still is a dog. Two dogs cannot produce a cat, a canary, or a cow. Animals of the same species and kind repeatedly reproduce offspring that have minor differences in appearance. There are absolutely no fossil records that support some intermediate form between and cat and a dog. Certainly, no one has observed this happening at any point in history. Here is the test for a true science project to become accurate and accepted in the scientific world. (1) True science must have clear evidence that the process can be proven in a scientific laboratory, (2) or that someone actually observed the animal changing from one kind to another. Evolution offers neither of these.

Here are some of the incredible changes that evolutionists claim have been produced: small microscopic sea organisms evolved into insects and fish, fish evolved into birds, birds evolved into reptiles, reptiles into all types of animals and prehistoric creatures, the family of apes and monkeys evolved into man. According to these supporters of evolution most of the original animal types died out and became extinct, including the intermediate forms. In this line of thought or imagination the stronger animals survived and those that were weak were likely killed off by their predators. Thus, the trail of evolution becomes a fairy tale left to man's imagination.

Evolutionists claim that life is still in the process of evolving into a gradual upward complex structure. The truth is that the universal laws advocate that matter left upon its own begins to break down rather than climb to some higher level of complexity. Things do not improve with age, but rather just the opposite. One only needs to look around and observe this in nature. If things are left to themselves, they will always break down, self-destruct or become less than the original product, whether it is a building or something with life blood cells. Even nature contradicts this ridiculous theory that creatures turn into others kinds of creatures.

Another big objection to the theory of evolution is the fact that the strongest creatures or most aggressive ones would inherit the earth as things around them continued in an upward direction. The truth is all around which proves that the world is actually in a state of degeneration rather than in some upward progression. The biblical record clearly demonstrates that this pattern was set in motion in **Genesis 3**, when man's sin stained life caused his fall and placed him under the curse of God. It was in the same chapter in Genesis that the Lord began to unfold His eternal plan of redemption for mankind as He pronounced their punishment. This proved His love for man, which gave them the hope of being restored.

The Bible teaches that God's original creation was perfect, without any flaws. Following that line of thinking, it is easy to understand that perfection cannot be improved upon and if things changed, then it would be a downhill trend. That is precisely what occurred after the fall of man in the Garden of Eden. Thus, the Great Flood was a direct result of the sin and perversion of the ancient world. The God of Creation had to bring His judgment upon the depraved world of people that turned their backs to a God that loved them. God's judgment will ultimately fall upon any group of people that move away from the God, whom they refused to believe and obey. Obedience brings God's blessings upon His people, but disobedience will bring about the wrath of God.

When God began to quiz Job, which is considered by most major Bible historians to be the oldest Bible book written, God begins with questions about the foundation of the earth. Please listen to the questions that the Lord used in speaking to Job about creation. The Lord said to Job, *"Where were you when I established the earth? Tell me if you have understanding, who fixed its dimensions? Certainly you know who stretched a measuring line across it? What supports its foundations? Or who laid its cornerstone while the morning stars sang together and all the sons of God shouted for joy? Who closed the seas behind doors when it burst from the womb, when I made the clouds its garment and thick darkness its blanket, when I determined its boundaries and put its bars and doors in place, when I declared: "You may come this far, but no farther; your proud waves stop here"?* **(Job 38:4–11).**

In the continuing discourse, the Lord says, "*Have you traveled to the sources of the sea or walked in the depths of the ocean?*" **(Job 38:16).** God continues to question Job with these difficult questions, *"Have the gates of death been revealed to you?"* **(v. 13)** *"Have you comprehended the extent of the earth? Tell Me, if you know all this"* **(v. 18).** *"Where is the road to the home of light?"* These are obviously questions that no man can answer, since man has no real understanding of all that God did in creation.

In the **Psalms 50,** God declares that *"every beast of the forest is Mine, and the cattle on a thousand hills. I know all the birds of the mountains and the wild beasts of the field are Mine"* **(v. 10–11).** God declares that He has not only created all things, but that all things belong to Him. After the Lord quizzes him, Job says, "*Therefore I have uttered what I did not understand, things too wonderful for me, which I did not know.*" It appears that Job has no depth of understanding of God, yet he is the greatest of all of God's servants in that day **(Job 1:8).** The Lord says to Satan, *"there is none like him on the earth, a blameless and upright man, one who fears God and shuns evil?"* Isaiah said it well, when he declared, "*For my thoughts are not your thoughts, nor are your ways My ways, says the Lord. For as the heavens are higher than the earth, so are My ways higher than your ways, and My thoughts than your thoughts"* **(Isa. 55:8–9).**

There is no doubt that man is not on the same level with that level of the Lord God. Man does not have the knowledge, the understanding nor the ways of even to begin to comprehend the things about the Lord. The only possible way that one will discover those mysteries of God is that God will have to reveal Himself to lowly man. Thus, it takes revelation from the Higher Being for the lesser to understand Him and His ways. It is clear from biblical teachings that God knows all things about His people and His creation, which is discussed in another chapter about the omniscience of the Lord. Whereas Job was a great man on the earth in his day and he did not know the secret things about God that were revealed during his interrogation. How much does this generation know about the Creator? If man would be truthful, then he would confess that he does not even know who that greatest man of God is on earth this day.

Perhaps, man is not looking to God for the answers of his origin, but believing and trusting in others to inform him. Certainly the world says they have all the answers about the origin of life, yet the truth is the world has strayed away from the One source that they need in revelation of the Truth, which is revealed in the Bible. The Scripture teaches that if man discovers the Source of Life, he will find the origin of all life with purpose in the One Supreme Being. Apart from the Lord God, man will follow endless paths of unrighteousness and hopelessness. Hope is all about the future and the future of man, like his past, is found in Jesus Christ, the Righteous One. The world searches endlessly in all the wrong places.

It is interesting to note that practically all those countries under communist leadership, as well as, those that have come out of communism teach the theory of evolution without a Divine Creator. There is virtually no part of academia today that has not been touched, and molded, by the belief in evolution. As the citadel of evolution is crumbling all about, it's interesting that evolutionists are more and more commonly crying that it is a fact **(20).** Those that believe in Christ should not be overwhelmed with fear that evolution will prevail, since falsehood cannot continue over extended periods of time without being found out. God is more grieved by powerless Christians than He is over powerful atheists **(21).**

If any student of the Bible or science would view a single cell under a microscope and study the complexities of the ingenious nature, he would conclude that a Master Architect had to have placed all the parts together. Molecular biologist, Dr. Michael Denton argues, *"Perhaps in no other area of modern biology is the challenge to evolution posed by the extreme complexity and ingenuity of biological adaptations more apparent than in the fascinating new molecular world of the cell"* **(22).**

It is a proven scientific fact that a single drop of ditch water can hold 500 million microscopic creatures so small that a teaspoon of water would be to them what the Atlantic Ocean is to us **(23).** These appear in thousands of different species living together, but having their own particular characteristics that make them different from each other. With all these complexities of a single cell, how can anyone argue that these systems just line up by chance. A mere piece of skin the size of a postage stamp requires three million cells; a yard of blood vessels, four yards of nerves; one hundred sweat glands, fifteen oil glands, and twenty-five nerve endings! **(24).** Yet the evolutionist demands one to believe that blind forces of chance produced complex human beings. It would be much easier to believe that Webster's unabridged dictionary was produced from an explosion in a printing shop.

Every living human cell contains DNA (deoxyribonucleic acid) molecules, which determine all the characteristics of a person right down to fingerprints, hair color, skin color, eyes, height, arrangement of 206 bones, 600 muscles, 10,000 auditory nerve fibers, 2 million optic nerve fibers, 100 billion nerve cells, 400 billion feet of blood vessels and capillaries, which could be held in the end of a teaspoon **(25).** All the complexities of design certainly point to a super intelligent designer, rather than blind chance.

EVIDENCE THAT OPPOSES THE EVOLUTION THEORY
(A Summary of 5 Facts Sheet)
by Dr. Charles Pratt

Christians and others who oppose the Theory of Evolution will always need plenty of ***FACTS*** for the opposing forces to consider, which have been tested by ***TRUE SCIENCE***. Here are some ***CONSIDERATIONS*** that hold merit against those false assumptions. Below is a **quick summary with 5 facts** that may be used by those that oppose the theory of evolution:

1. One issue that evolutionary theory followers use is that there was one little cell or period that spun around in the primordial soup in the swamp. That little period got energy, yet they do not attempt to say how that it received that energy. Those that follow this line of thought do not have an answer, so consequently, they quickly move their thinking beyond any explanation. True science always explains where energy comes from to begin any process of true science.
2. Another popular teaching of evolutionists is that the earth is billions of years old, which greatly conflicts with the Word of God. Any Bible student can put forth the teachings of the incredible, intelligent Creator. He created intel-

ligent human beings from the beginning that were created in the image of the Master Designer and Creator.

There were absolutely no defects in the mind of Adam nor any early man, since they were all made in the very image of God. Moses, a highly educated man of God was given an assignment from God to write a huge portion of the Old Testament. He received from God the exact number of years that those first human beings lived and carefully recorded these with exactness in the first five books of the Bible. It is easy for intelligent man to add the years in Genesis 5, Genesis 10, other places, such as the length of time that Israel spent in Egypt, or the period of years of every King that served in Israel. The exact time frame of every king is recorded in the Bible. All of these time periods may be added together, which are slightly over 6000 years and nowhere close to the billions of year theory stated by those which oppose the biblical views.

These materials produce a total of 1656 years from creation to the Great Flood of Noah's day, with nearly 2400 years from the Flood to the birth of Jesus. In conclusion, there were approximately 4000 years from the creation of the earth and man to the coming of the Messiah.

3. Here are a few questions that any curious student of the Scriptures may want to ask those that are well versed in God's Holy Word. Perhaps, those students would be willing to ask themselves these questions before asking others such questions. Does the student of the Bible trust the intelligence of their forefathers that recorded all of these dates? Can a person trust the Bible dates as authentic, God given information that He gave to over 40 different human writers as they wrote the biblical records?

   Every Bible student should point out that those highly intelligent people were the ones that gave us the dating system for the earth, which is centered upon the coming of Jesus Christ. They are the ones that knew of the centrality of the universe which would be centered around the coming of Jesus and His redemptive work at the Cross of Calvary. They approved the calendar with the label of **B.C.** *(Before Christ)* time frame. Further, they labeled the years that followed as **A.D.** *(After Death of Christ)* period of time. This alone should speak volumes to those seeking for the Truth. The BIG question is why would those leaders choose the death of Jesus Christ as the central event for all of history of the world? Were they deceived or simply being guided by the Lord in their recordings? Remember these people were made in the image of God, thus they were not a bunch of dummies.

Some age old questions arise as to the length of a day, which may not have been 24 hours of time. Those great arguments continue today whether a day or a year contain the exact same time frame as an ancient day or a year compared to today. Once the Creator placed all the planets into orbit during the first week of creation, there is zero evidence or reference that God re-created anything differently at some later date. The truth is that God created time during the creation week without any changes as he placed all things into their proper place in each segment of creation. Since day 4 when the Creator put everything into motion rotating around the sun, He has not changed one single thing. According to Genesis 8:22, time has not changed, nor will time ever change as long as the earth remains.

4. One huge piece of evidence against evolutionary theory is the fact that the planets have motion but they are not running down as would be the case had some planets crashed into another planet. Over billions of years that energy level would have diminished to some extent, if not have become nil. The facts are clear that God hung these planets into outer space and they all rotate in various directions with no reported crashes. The earth spins on its axes over every 24 hours and rotates completely around the sun every 365 ¼ days continuously for over 6,000 years. The Architect of

the Ages placed all things into their place when he spoke the worlds and planets into existence. These planets cannot get out of their rotations, unless the Manufacturer of all planets grants them His permission. As an old song goes, "He's got the whole world in His Hands."

5. If the planets had been around for billions of years, then there would be dust very deep upon each of these planets and no person or creature could possibly navigate their way around with ease. This was proven at the point of the first men and spacecraft that landed on the moon on July 20, 1969. Those that developed this magnificent spaceship were certain that the dust would be dozens of feet deep on the surface of the moon. As a result of this thinking, they fabricated long extended metal legs to come out of the spacecraft so to keep it safe above the dust when they landed. However, upon landing on the moon, they discovered only a very few inches of dust was on the surface of the moon, which indicated a young moon. The same likely applies to all the planets.

Theistic Evolution Theory

There is an increasing popular belief today that combines the creation story of Scripture with the chance evolutionary events of science often called "Theistic Evolution," which attempts to reconcile these views. Unfortunately, there are many serious problems for the student of science, as well as, the Bible student that cannot be explained with this view. This belief often accepts the beginning of life with a one cell organism that evolved upwards into higher forms of life to man, with God's watchcare or blessings. The major problem is that this story contradicts the biblical account of man and animal life being created as full grown species. Many people that do not want trouble or confrontation are willing to make a compromise with the opposite side and try to marry these two opposite points. Evolutionists rarely want to agree with Christians on anything, but some are willing to at this point, since this belief proclaims that the first eleven chapters of the Bible become an allegory, instead of the factual story of how all things were created by God **(26)**.

Many people who claim to be Christian and seem to be genuinely Spirit filled followers of biblical teachings say there is no conflict with the creation account in the Bible and Theistic Evolution. These will often claim that God seems to have created and has permitted that creation to continue with some occasional interventions **(27)**. Whereas, over a generation of a single life that one person may carry on his business, his church work, family connection and then when he dies, another comes along and makes changes in all of the categories. They contend that God seems to approve of some of these changes. This sets up a pattern of God working with the human race through His fixed laws.

All of these teachings seem to blend with things over a long process of time. However, the Bible teaches that God is not standing on the sidelines. Since God walked with Adam and Eve regularly, likely daily, while He fellowshipped with them in the Garden. Soon after that first sin in the Garden, God sought the first family rather than leaving them in their sinful condition. This would indicate that if a person that belongs to the Lord commits sin, he may expect the Lord God to show up very soon thereafter. He takes the initiative, since the Lord loves every human being that He is unwilling to leave any person in that condition, thus He seeks sinful men. This is the truth about a loving Savior that is found all through the Bible.

Unfortunately, Pope John Paul II, leader of the Roman Catholic Church in a speech in 1996 endorsed this belief of theistic evolution. However, he did indicate that he held a few reservations concerning the view. Then his successor, Pope Francis in 2014, mentioned that this was "a possible process" that the Lord could have used in creation. Both of these Christian world leaders may have seriously damaged the biblical account in a serious manner for years to come. Theistic Evolution potentially takes away God's glory. It strips Him of His majesty and power, which are the very characteristics that define Him. God created the original world out of His omnipotence, His glory, and His intrinsic, total perfection, which never changes **(28)**.

There are many other regular church attending people today that have long accepted this view, rather than giving diligent study of the facts in science and creation. In fact, a growing number of (so-called) conservative scholars embrace theistic evolution to be divinely designed and governed. They believe that Genesis is speaking more of the relationship between God and creation than in presenting a scientific or historical explanation of how creation actually occurred. There is absolutely no way that theistic evolution can fit into the biblical account of creation. It is a mere compromise for the uninformed person that does not understand the tenets of faith of evolution nor creationism, since they are like gasoline and water that will not mix. The Genesis record of creation and the story of evolution are absolutely incompatible, yet there are so-called students of the Bible that are attempting to mesh them into the same cake. It is simply a fabricated theory for those that desire to be accepted by the world.

One reformed theistic evolutionist is Jobe Martin, who has written on this subject and explains how that he was transformed in his thinking regarding evolution. Martin in his book entitled **"The Evolution of a Creationist"** explains how that he was challenged by 2 of his students one day regarding his beliefs on creationism. His views shifted when Martin accepted Jesus as his Savior from an agnostic evolutionist to that of a theistic evolutionist **(29).** By his testimony, Martin, an Air Force dentist during the height of the Vietnam War, wanted to make sure that his thinking was right by offering a prayer to God and testing Him, if He was real. His prayer to God was, "God, if you are up there, you have two choices. Either you can show me the girl I am going to marry, or you will see the wildest Air Force officer you have ever seen." After that prayer, Martin had determined to go out and really live it up.

However, God did take that challenge and that very day, he met the girl that he would marry. That next day, Martin told that girl that he was going to marry her. That Sunday, he began to attend a church and was introduced to Jesus as his Savior by the pastor of that church. WOW! According to this man's testimony, he knew that God took his challenge.

Truly the Bible is not a science book, although it does contain much scientific truth, while most science books attempt to supply knowledge that can fill the head with even more questions that will never satisfy the heart of a seeking man. It was obvious that Jobe Martin was seeking answers that science could not deliver. There will always be conflict between the major voice of science and scripture. The fact is that each Believer must accept by faith which belief to adopt. Faith is the basis for a belief in any system of life. Since this is true, the matter of belief in a scientific creation story becomes a religion to that advocate. He accepts it as his belief of creation and will often oppose the Divine creation account.

Evolution can be thought of as sort of a magical religion. Magic is simply an effect without a cause, or at least a competent cause. "Chance," or "time," and "nature" are the small gods enshrined at evolutionary temples. Yet these gods cannot explain the origin of life. These gods are impotent. Thus, evolution is left without competent cause and is, therefore, only a magical explanation for the existence of life **(30).**

D. Gap Theory

Another popular view of creation that involves a great following of believers is the "Gap Theory" or sometimes called the "Big Gap Theory." Those that believe this theory contend that a gap of years exist between Genesis 1:1 and 1:2. The contention is that there were possibly millions of years between these verses which explains the theory for those supporters that say the earth is billions of years old. Others use this argument for those that believe God used the process of evolution during the millions of years to produce man. There is no concrete proof of this view that these years ever existed. Perhaps, this theory, which cannot be verified as a fact is often followed to get Satan into the picture. Prominent theologians have written widely promoting this particular viewpoint such as Thomas Chalmers, G. H. Pember, and Arthur Pink. These point to an ancient world with possibly millions or billions of years between the first two verses of the Bible. They believe that Satan was in the original world in verse one as a part of God's creation. Lucifer ruled the world, but committed sin, so God judged the old world with Lucifer, as a result the earth became formless, desolate mass that became **Genesis 1:2 (31).**

This theory took place in pre-Adam times, in connection, perhaps, with another race of beings, and consequently, does not at present concern us **(32).** This theory often gives rise to the idea of bringing in the prehistoric creatures that roamed the original earth with teachings on the fall of Satan and his demons. Many of these contend that this makes an accommodation for the present-day arguments for the huge number of missing years. It also accommodates the teachings of a historical world where large size animals roamed the original earth, which God remanufactured from the ruins. The bones of these so-called prehistoric animals fit into this theory. The Bible clearly teaches that God made the "great" size animals on day number 6 prior to His creation of the first man **(Gen. 1:21).**

In fact, the Bible actually disproves the teachings of these pre-historic mammoths living prior to man. Again consider the book of Job as the oldest writing in the Bible and in this account near to the end how God begins to question Job by asking him, "Look at Behemoth, which I made along with you; he eats grass like an ox. See now, his strength is in his hips and his power is in his stomach muscles. He moves his tail like a cedar" **(Job 40:15–24)**. This was a very huge creature that man could not tame. God asked Job, do you see him, which clearly indicated that man and the giant creatures lived in the same world and not a different world. Job clearly lived in the same world with these huge animals. God continues by calling Job's attention to even more fierce animal that the Lord describes. "Can you draw out Leviathan with a hook, or snare his tongue with a line which you lower? **(Job 41:1–34).** This huge mammoth breathes out fire from his nostrils and smoke, which reads like a dragon. He is fearless of all beings with such enormous size that he can boil up the waters in the ocean **(v. 31).** These, as well as, other huge size animals lived on the earth at the same time of Job and not in some previous world.

Such beliefs compromise the biblical account by injecting unproven data into the story of creation. Again, no true scientific evidence is available to fully support such theories. "Belief in the Gap Theory spread widely after its inclusion as a footnote in the Scofield Bible in the 1917 edition. The Bible says that no death existed before Adam's sin and God's curse, therefore this theory is nonbiblical" **(33).**

This view presents a host of conflicts with the Bible for those that believe in the Big Gap Theory. The first problem is that if Satan ruled over a kingdom between Genesis 1:1 and verse 2 of this chapter means that sin had already been in the world long before the Garden of Eden was created. Then sin was present prior to the creation of the first man that presents a whole host of problems for students of the Word of God. God's creation had to have been imperfect rather than as the Lord declared that all things were good, which means perfect. After the creation of man, God said it was very good, which would not be accurate if the devil was present during the creation. The previous world that had been destroyed would have been wicked and maybe violent with death all over the place. God would have been attempting a re-make of the fallen world, while Satan was casting his dark shadow over the new creation. If suffering and death existed for billions of years before the sin of Satan, then the subsequent sin was prior to Adam **(34).** However, the Bible teaches a perfect creation with Adam and Eve committing the first sin on earth. Thus it is impossible for a Believer of the Bible to embrace the Big Gap theory.

Intelligent Design Theory (ID)

A more modern view of creation that is being considered today is that of "Intelligent Design" or often known as ID. This view somewhat falls short of biblical creation account, but is founded on scientific facts that point toward a purposeful design rather than blind chance. This is the closest to the Bible view, so that many Bible students are completing this view by stating that God is the Intelligent Designer as opposed to random chance taught in evolution.

The ID view suggests that a student of science observing the vast complexity of a living cell will quickly conclude that there is no possible way that all things can simply line up to form a complex cell. Scientists have reported today that there are as many as 100 trillion cells in the human body performing thousands of specific functions and each cell contains about a trillion atoms (**35**). Again, the human body is so complicated and complex, there is almost no chance in multiple trillions that this creature could have happened by mere chance. This means that evolution is not based upon true science. There seems to be a growing following of this belief today. In fact, there is a movement underway for schools to teach this belief alongside of the theory of evolution. Intelligent Design would create the opportunity for students to form their own opinion about the identity of this Designer.

A proponent of ID is a highly respected scientist and college professor from Lehigh University name Michael Behe. Dr Behe, whose work includes a 1996 bestseller called **Darwin's Black Box** argues that ID should be taught in public schools. He concludes that evolution cannot fully explain the biological complexities of life, suggesting the work of an intelligent force. The ID belief stops short of naming the force behind the belief, although Behe is a Roman Catholic, who testifies that he believes it is God. Behe has written many other books and publications stating that the most serious error of Darwin's theory of evolution is the fact that there is no molecular level of cell complexity involved in Darwin's hypothesis. The jury is still out on whether this view will receive a fair opportunity to be heard by public school students.

Biblical Creation

Another popular view held by Christians, Jews and many other non-Christians is that the Bible story of creation is an accurate version that God has granted to man. This study seeks to be a guide to helping those who believe in the biblical creation account and help them to strengthen their position for a Divine Creator. Perhaps, it will shine light upon some that have not yet formed an opinion or those that are searching for the truth.

The biblical view firmly establishes that Jesus Christ, man's Savior is the God who spoke the worlds into existence. In fact, in the mind of God, Jesus had already gone to the cross and had been crucified for the sins of the entire world, since the Bible speaks of the Lamb of God that was sacrificed before the foundation of the world was put into place **(Rev. 13:8).** Thus, the cross was of such significance to the God of creation that He worked out the process ahead of time, so that the cross actually precedes the act of creation. The cross was at the heart of creation, since the Christ of the cross was integrated into every aspect of God's creative work. The most prominent place is found following the first act of sin of man in Genesis 3.

This study deals with the cross of Christ in the first eleven chapters of the Bible. There are those who claim that this portion of Scripture is not an account of actual events that happened, but rather a writer's way of telling about the happenings through fables, allegories or metaphors that convey the principles of truths. The truth is in most writings the plain message is usually given before any hidden meaning. There is no evidence given in the Genesis creation account of this being an allegory, but rather a statement of fact.

The Bible declares very emphatically, *"Know that the Lord, He is God; It is He who has made us, and not we ourselves"* **(Ps. 100:3).** In fact, the Bible further states *"You covered me in my mother's womb. I will praise You, for I am fearfully and wonderfully made"* **(Ps. 139:13–14).**

Obviously, these views do not include all the various beliefs present in today's world. However, this study only mentions the more popular beliefs, but centers upon an intensive study of the creation account from the Scriptures that places the cross of Christ at the center of creation. With such a broad belief system as the aforementioned views alone, one can sense the battlefield where two of the great authorities of life meet: the authority of Scripture and the authority of science **(36).** This study does not attempt to reconcile these two giants, but rather shows how true science does not conflict with the Word of God. Much that is called science today, simply operates from a very unscientific basis that makes assumptions without looking for answers. The power and presence of the Holy Spirit is always the teacher to shine light upon the dark or unenlightened mind of the human race that sincerely seeks the truth.

One of the strongest arguments for biblical Creation is based upon the complexity in creation, especially with human bodies, animal bodies and plant structure. The complex makeup of all living beings could not have evolved or come into existence by blind random chance as evolutionists have suggested. In fact, the molecular makeup of a tiny raindrop or a living cell are far too complicated to reproduce left to chance. Consider just one cubit inch of rain water and its composition.

Empirical or true science definitely point to some Supernatural involvement in the whole detailed process. There is no way that "the living" part could be possible without the involvement of the Supernatural.

Another strong argument for the biblical Creation account is that there is no one on earth that can create something out of nothing, such as the case in the biblical narrative. God came out from nowhere, stood upon nothing and literally spoke all things into existence is taught in the Bible account. This may not be defined as logical; however there is no other evidence that creates any sensible solution to the question of "How did all of this come to exist?" A more honest question would be "Who made all of this world and the things in it?"

Chapter 13

Various Views of Creation Expanded

1 Flamming, Peter James, ***Layman's Library of Christian Doctrine***, Broadman Press, Nashville, Tennessee: 1978, pg. 10.
2 Lisle, Jason, ***Why Genesis Matters***, Christian Doctrine and the Creation Account, The Institute for Creation Research, Dallas, Texas: 2012, pg. 10.
3 Barnes, ***Barnes Notes***, pg. 28.
4 Stanley, Charles, ***Enter His Gates***, A Daily Journal into the Master's Presence, Thomas Nelson Publishers, Nashville, Tennessee: 1998, pg. 109.
5 Johnson, P., ***Defeating Darwinism by Open Minds***, InterVarsity Press, Downers Grove, Illinois: 1997, pg. 15.
6 Morris, ***The Book of Beginnings***, Vol. 1, pg. 31.
7 Ham, ***The Genesis Solution***, pg. 45.
8 Ralph O. Muncaster, ***Dismantling Evolution*****,** Building the Case For Intelligent Design, Harvest House Publishers, Eugene, Oregon: 2003, pg. 23.
9 Behe, Michael J., ***Darwin's Black Box***, Touchstone Book published by Simon & Schuster, New York, NY: 1996, pg. 78.
10 Boice, James Montgomery, ***An Expositional Commentary***, Genesis, Vol. 1, Published by Baker Books, Grand Rapids, Michigan: 1982, pg. 42
11 ***Ibid.*** pg. 42.
12 Darwin, ***On The Origin of the Species***, pg. 201.
13 ***Ibid.***
14 News Article, ***The Jackson Sun***, What happened to Madeline Murray O'Hair, Mar. 13, 1998.
15 ***The Genesis Solution*****,** pg. 14.
16 Morris, Henry M. III, ***The Book of Beginnings***, Institute of Creation Research, Vol. I, Creation, The Fall and the First Age, Dallas, TX: 2012, Vol. 1, pg. 8.
17 Huse, Scott M., ***The Collapse of Evolution***, 3rd Edition, Baker Book Company. Grand Rapids, Michigan: 1997, pg. 84.

18 ***Ibid.***, pg. 46
19 Rhodes, Ron, ***Creation vs. Evolution Debate***, Harvest House Publishers, Eugene, Oregon: 2004, pg. 76–77.
20 Kennedy, D. J., ***Origins: Creation or Evolution?***, Coral Ridge Ministries, Fort Lauderdale, Florida: 1986, pg. 2.
21 Boom, Corrie Ten, ***Each New Day***, A Daily Devotional Study, Fleming H. Revell, a Division of Baker Book House, Grand Rapids, Michigan: 1977, pg. 80.
22 Denton, Michael, ***Evolution: A Theory in Crisis***, Alder and Adler, Bethesda, Maryland: 1986, pg. 23.
23 Phillips, ***Exploring Genesis***, pg. 44.
24 ***Ibid.***, pg. 49–50.
25 ***Evolution: A Theory in Crisis***, pg. 117.
26 Rhodes, ***Creation vs. Evolution Debate***, pg. 58.
27 Boice, James Montgomery, ***An Exposition Commentary***, Genesis Vol. I. Baker Book House, Grand Rapids, Michigan: 1982, pg. 53.
28 Ham, ***The Genesis Solution***, pg. 67.
29 Martin, Jobe, ***The Evolution of a Creationist***, Biblical Discipleship Publishers, Rockwall, Texas: 1994, pg. 10
30 Wysong, Randy L., ***The Creation/Evolution Controversy***, Inquiry Press, East Lansing, Michigan: 1976, pg. 63.
31 Boice, ***An Exposition Commentary***, Vol. 1, pg. 57.
32 ***Ibid.***
33 Ham, ***The Genesis Solution***, pg. 61–62.
34 Morris, ***The Genesis Record***. pg. 47.
35 Muncaster, ***Dismantling Evolution***, pg. 122.
36 ***Layman Library of Christian Doctrine***, pg. 11.

CHAPTER 14

A CONNECTION TO THE CROSS

As a follower of Jesus Christ and a student of the Word of God, it is important that each true Believer not only know what happened in the Beginning, but how it is connected to the experience of Jesus upon the Cross. From **Genesis 3:15**, God's Word reveals a plan for every person to be brought back to the Lord, even though all have sinned and come short of His marvelous plan to live the righteous life on earth. Every human being has done wrong and broken at least one of the Laws of God. The only perfect person that ever lived was Jesus Christ, the God-Man.

It is noteworthy to mention that Man fell in the garden of paradise, which was literally a perfect environment with not one single flaw. If man could fall in a perfect setting, then he will never come anywhere close to a perfect walk with the Lord in this fallen world today. None of this caught our Lord off guard or by surprise. The Bible declares that He knows what is in the heart of man and even the thoughts of every man **(Luke 6:8).** Again, in **Acts 1:24**, the Bible states that God knows the hearts of all men. Man will never be able to correct his past departures from the straight and narrow path that the Lord called each person to walk. In God's infinite foreknowledge, He planned for man's course of correction through His perfect Son, Jesus Christ as the only way to heaven.

This course of correction would ultimately cost the Lord everything, yet He was willing to pay the supreme price to give sinful man a plan of redemption. That expensive plan would place the Cross of Jesus right in the midst of pardoning mankind. The story in Genesis to correct the first man Adam involved **redemption** through the blood of an innocent animal **(Gen. 3:21**). That particular animal that gave its life for a covering for sinful man is not revealed in this verse in Genesis, but shortly is revealed throughout the Old Testament as a lamb. That lamb was the temporary prototype that would be revealed as Jesus Christ in the New Testament by John the Baptist during the baptism of Jesus **(John 1:29).** In fact, He is the perfect Lamb of God. This Lamb in the New Testament would not just cover the sin, but would take away the sin, according to John the Baptist.

The Lord presented the Old Testament people a clear version of the story of redemption in Genesis 22, when God called Abraham to sacrifice his son Isaac on Mt. Moriah. As these two were walking up that mountain, Isaac asked Abraham a question. Isaac said Daddy, we have the fire and the wood but "where is the lamb?" Abraham said, "My son, God will provide for Himself the lamb for the burnt offering" (22:8). A few thousands of years later, God did place His Son, Jesus upon a cruel Roman cross to die for every human being on that very same mountain, which was called Mt Calvary in the New Testament.

In order for man's redemption to be fully paid, there had to be the cruel cross in the midst of the cleansing process with the shedding of the Blood of the acceptable Lamb of God. Jesus was the only One that could satisfy that payment. The stem of the cross becomes the staff of life, and in the midst of the world life is set up anew upon the cursed ground **(1).** Later the writer John would declare that Jesus became the propitiation for the sin of man **(1 John 2:2).** In some translations that word propitiation is translated as an "advocate" perhaps something like an attorney to speak for someone who is guilty. Jesus is that one who will stand before the Great Judge and say this one belongs to me or he has my blood upon his life.

The Hebrews believed that the blood was synonymous with "life." A life (or blood) had to be given, before the transaction would become complete. The Hebrews' writer also stated that without the shedding of blood, there is no forgiveness of sin **(Heb. 9:3).** It is true that Jesus saw Himself as the Good Shepherd **(John 10:11)** that would willingly lay down His life for the lives of all mankind. It was not possible that the blood of bulls and goats could take away sins **(Heb. 10:4).** It is abundantly clear that the life of the Good Shepherd would be the only One that could rescue a sinner from the flames of Hell, which all men deserve. By the grace of this loving Shepherd, man was offered a place in Heaven rather than Hell, providing sinful man would truly repent of his sin and invite the Good Shepherd to become the Master of his life.

So the cross was absolutely essential for everyone who would choose to accept God's plan of redemption. Every man must come to the awareness of God's plan, before he can be delivered from his predicament of facing an eternity separated from the loving God who created each person.

One of the big problems of getting people to the point of redemption today is the fact that most folks need to know that they are hopelessly lost without a relationship with the Good Shepherd. Vance Havner, a great evangelist of the twentieth century often stated the greatest problem of getting people saved is first getting people lost. He declared that people seldom see themselves lost and going to hell without a Savior. Truly, it will take every redeemed person to make the connection with those that are not aware of their spiritual condition and share that beautiful story of the cross and the resurrection power of Jesus, our Good Shepherd. If one is not certain of the destiny of a friend or loved one, please share this Good News story of the hope that is in Jesus Christ.

Chapter 14

The Connection to the Cross

1 Bonhoffer, ***Creation & the Fall***, pg. 93.

A SPECIAL BLESSING FOR ALL MANKIND

In the process of bringing this section of this book to a conclusion, my best friend in this world made the following suggestion to me regarding any person reading this conclusion. That best Friend is Jesus and early one morning, while finishing this book, He placed this message on my mind. So I am sharing it with whoever has chosen to read this.

This story has been shared around the globe on at least five continents and dozens of nations about **how to get to Heaven from planet earth, without a spaceship.** Read these instructions of how **whoever** reads this can do it correctly.

HUMBLE YOURSELF. Just bow your head in reverence to the Holy God that spoke everything into existence, since HE is above all people at all times. From a point of respect, since no person can look upon HIS face from earth, please bow your head to shut out the things of this life.

CONFESS THAT YOU ARE A SINNER. The Bible clearly states that every person has committed sin and that includes you and me. It further declares that there are NO exceptions, since all have lied, cheated, stolen, said unkind words, disobeyed parents or thought upon bad things that we could say to another human being. All of these and more are sins against a Holy God that made us with dust of the ground. Simply tell HIM who is constantly listening that you are a sinner and you know that you, like everyone else, has done wrong.

ASK FOR GOD'S FORGIVENESS OF YOUR SINS. Simply say, Lord Jesus, please forgive me of all my sins and cleanse me, so I may be set **FREE** from the penalty or debt of sins. If you are sincere with this request, then He will **PARDON** you of all that you have ever done wrong throughout your life.

ASK JESUS TO ENTER YOUR LIFE AND FOR HIS FREE GIFT. He (the Holy Spirit) will quickly enter your heart and **CANCEL YOUR DEBT OF SIN** with a full pardon and write your name down in Heaven at the same time. At the very same time Jesus enters into you He will place your name in Heaven in the Lamb's Book of Life, where it will never be taken out. No one can take it away from you, since it will be written in the **BLOOD OF JESUS**. This means that you will be "Born Again" and will never ever go to hell.

KNOW THAT YOU ARE GOING TO HEAVEN. The Bible is very clear that once whoever sincerely prays and truly means it will have a fully paid first-class **TICKET** to Heaven. It is not by works or making a purchase of giving from your materials, but by **HUMBLING YOURSELF** before the Creator of this vast universe, **CONFESSING** that you are a sinner and asking for His forgiveness of the **SIN DEBT** that no one can **PARDON**, but Jesus Christ. He went to the **CROSS** for your **SIN DEBT**, as well as, all others that come unto Him repenting of your sins. Jesus wants every human being on earth to do this before they die, so Heaven is your eternal home when you die. All mankind will die and then all will be judged ready for Heaven or sentenced to go to hell. Please remember this promise from His Word of Truth: "Whoever calls upon the name of the Lord shall be saved." (Rom. 10:13).

SELECTED BIBLIOGRAPHY

Chapter 1: Introduction to the Study of Genesis

1. **www.Gallup.com, Gallup Poll Survey,** Views of Origin of Human Beings, May Survey Study, 6-1-12.
2. Live Science, ATechMediaNetworkCompany, ***Science News***, March 8, 2017.
3. **www.Gallup.com**, **Gallup Poll Survey.**
4. ***AFA Journal,*** American Family Association, US Evangelicals Confused About Theology, June 2017 issue, pg. 8.
5. Ham, Ken, ***The Lie: Evolution***, Genesis, the Key to Defending Your Faith, Master Books, Inc., 23rd printing, Green Forest, Arkansas: 2002, pg. 17.
6. Huse, Scott M., ***The Collapse of Evolution***, 3rd Edition, Baker House, Third Edition, Grand Rapids, Michigan: 1997, pg. 17.
7. Ham, ***The Lie: Evolution,*** pg. 71.
8. Ham, Ken & Taylor, Paul, ***The Genesis Solution***, Baker Book House, Grand Rapids, Michigan: 2000, pg. 45.
9. Thieme, R. B., Jr**, *Creation, Chaos, and the Restoration,*** R. B. Thieme Bible Ministries 1995, pg. 1.
10. Morris, Henry M. III, ***Book of Beginnings***, Vol. 1, Creation, The Fall, and the First Age, Institute of Creation Research, Dallas, Texas: 2012, pg. 13.
11. Ham, ***The Genesis Solution***, pg. 49.
12. Morris, ***Book of Beginnings***, pg. 9.
13. Darwin, Charles, ***The Origin of Species***, New American Library, a Division of Penguin Books, Ltd., 80 Strand, London, England: 1958, pg. 18.
14. ***Ibid.***, pg. 19.

15. ***Ibid.***, pg. 107.
16. Spence, H. D. M. and Exell, Joseph S., Editors, ***The Pulpit Commentary***, Vol. 1, Hendrickson, Publishers, Peabody, Massachusetts: pg. 7.
17. Menton, David, ***Apemen***, "Separating Fact from Fiction," Answers in Genesis, Petersburg, KY., 2010, pg. 13.
18. ***Ibid***, pg. 14.

Chapter 2: Title and Authorship

1. Barnes, Albert, ***Barnes Notes of the Old and New Testaments,*** Exposition of Genesis, Vol. 1, Baker Book House, Grand Rapids, Michigan: 1981, pg. 5.
2. ***Ibid.***, pg. 36.

Chapter 3: Frequently Asked Questions

1. ***Acts & Facts Magazine,*** **Institute of Creation Research,** Dallas, Texas, vol. 47, by Brian Thomas, from Thomas, B. Best, posted on Creation Science Updated from Dec. 2017, June 2018, pg. 20.
2. Morris, Henry II, ***The Genesis Record***, Baker Book House Company, Grand Rapids, Michigan: 1976, pg. 44.
3. Morris, Henry III, ***Acts & Facts Magazine***, ICR, July 2016, pg. 6.
4. American Family Association, ***AFA Journal***, Tupelo, Miss., September 2017, pg. 8.
5. ***Ibid***.
6. ***Ibid.,*** March 2018, pg. 10.
7. ***Ibid.***
8. Hebert, Jake, ***Acts & Facts Magazine***, The Bible Best Explains the Ice Age, November 2018, pg. 10.
9. ***Ibid.***, pg. 11.
10. Holt, R. D., ***Acts & Facts Magazine***, (ICR), Evidence for the Late Cainozoic Flood/Post Flood Boundary, Dallas, Texas: August 2016, pg. 9.
11. Wicander, R.& Monroe, J. S., ***Historical Geology***, 7th Edition, Brooks/Cole, Cengage & Learning, Belmont, California: 2013, pg.

12. Answers in Genesis, ***Ark Materials,*** Ken Ham, Williamstown, Ky.
13. ***IBID.***
14. Phillips, John, ***Exploring Genesis***, Loizeaux Brothers, Neptune, New Jersey: (originally printed by Moody Press), 1992, pg. 38.

Chapter 4: A Brief Overview of the Creation Theories

1. ***AFA Family Journal,*** American Family Association, December 2018, pg. 5. (pewrearch.org, 10/1/18).

Chapter 5: The Pre-Creation Account

1. Hodge, Bodie, ***Satan and the Serpent***, Answers in Genesis Series, A Biblical View of Satan and Evil, Hebron, Kentucky: 2014, pg. 10.
2. Thieme, ***Creation, Chaos and the Restoration***, pg. 3.
3. Hodge, ***Satan and the Serpent***, pg. 24.

Chapter 6: The Eternality of the Creator

1. Mohler, R. Albert, ***Decision Magazine***, Billy Graham Evangelistic Association, Charlotte, N. Carolina, December 2013, pg. 14–17.

Chapter 7: The First Creation Story

1. Morris II, ***The Genesis Record***, pg. 37
2. ***Ibid***, pg. 38.
3. Barnes, Albert, ***Barnes Notes of the Old and New Testaments,*** pg. 40.
4. ***Ibid.***, pg. 33.
5. Marsh, Spencer & Seelig, Heinz, ***Beginnings-Portrayal of the Creation***, Multnomah Press, Peabody, Maryland: 1981, pg. 19.
6. ***Ibid.***
7. Morris II, ***The Genesis Record***, pg. 59.

8. Ham, Ken, Snelling, Andrew, & Wieland, Carl, ***The Answers Book***, Answers in Genesis, Master Books, El Cajon, Calif: 1992, pg. 120.
9. ***SBC LIFE***, Southern Baptist Convention quarterly magazine, Nashville, Tenn., quote from ***US News and World Report***, January 1998 issue, pg.12.
10. Thieme, ***Creation, Chaos and the Restoration***, pg. 10.
11. Bradstreet, David & Rabey, Steve, ***Star Struck,*** Seeing the Creator in the Wonders of Our Comos, Zondervan Press, Grand Rapids, Michigan, 2016, pg. 208.
12. ***Ibid***.
13. ***Ibid***.
14. ***Acts & Facts Magazine***, Institute of Creation Science, October 2018, pg. 10.
15. Bradstreet, ***Star Struck***, pg. 207.
16. MacArthur, John, ***The Battle For The Beginning***, W Publishing Group, Division of Thomas Nelson, Nashville, Tennessee: 2001, pg. 106.
17. Bradstreet, ***Star Struck***, pg. 208
18. Phillips, John, ***Exploring Genesis***, pg. 44–45
19. Hovind, Kent, ***Creation Science Evangelism***, A resource supplement Creation Science, Pensacola: Florida: 2000, pg. 7.
20. Phillips, ***Exploring Genesis***, pg. 44.
21. Ham, ***The Answers Book***, pg. 21.
22. Gallings, A. P., ***Dinosaurs,*** A Pocket Guide, Answers in Genesis, Hebron, Kentucky: 2010, pg.
23. Horner, J. & Lessem, D., ***The Complete T Rex***, Simon and Schuster, New York, New York: 1993, pg. 18.
24. Morris II, ***The Genesis Record,*** pg. 68.
25. Ankerburg, John, pg. 5.
26. ***Ibid***., pg. 6.
27. Phillips, ***Exploring Genesis***, pg. 45.
28. Morris, Henry III, ***Science and the Bible***, Moody Press, Chicago, Illinois: 1951, pg. 35
29. ***Ibid***., pg. 41.

30. Hovind, Kent, ***Creation Science Evangelism***, pg. 37
31. ***Exploring Genesis***, pg. 41. pg. 47
32. Ross, Allen P., ***Creation & Blessings***, Baker Books, pg. 75

Chapter 8: The Second Creation Story

1. Bonhoeffer, Dietrech, ***Creation and Fall***, A Theological Interpretation of Genesis, McMillian Publishing Company, Inc., New York, NY: 1978, pg. 47.
2. Ibid., pg. 89.
3. Phillips, ***Exploring Genesis***, pg. 53.

Chapter 9: The Curse on the Earth

1. Phillips, ***Exploring Genesis***, pg. 56.

Chapter 10

1. Lisle, Jason, ***Why Genesis Matters***, Institute for Christian Research, Dallas, Texas: pg. 13.
2. **Phillips, *Exploring Genesis*, pg. 73.**
3. Stringer, Chris, ***National Geographic***, Global Journey, Natural History Museum, London: January 2013, pg. 48–49.
4. ***Ibid.***, pg. 49

Chapter 11: How the Flood Transformed the Earth

1. Ham, ***The Answer Book***, Vol. 1. pg. 137.
2. Phillips, ***Exploring Genesis***, pg. 79.
3. Ham, ***The Answers Book***, Vol 1. pg. 138.
4. ***Exploring Genesis***, pg. 85.
5. Morris, Henry, & Miller, William J., ***The Genesis Flood***, The Introduction to Historical Geology, 1952, pg. 156.
6. Ham, Ken, General Editor, ***The New Answers Book 1***, Answers in Genesis, Master Books, Green Forrest, Ark: 2011, pg. 136.

7. Morris, ***The Genesis Flood***, pg. 161.
8. ***AFA Journal***, The American Family Association, from Fox News, www.foxnews.com, 12/12/12, March 2013 issue, pg. 6.

Chapter 12: The Age of the Earth

1. Hovind, Ken, ***Dinosaurs Creation Evolution***, Creation Science Evangelism, Pensacola, Florida: 1990, pg. 5.
2. Morris, ***Acts & Facts Magazine***, July 2016, pg. 6
3. Thomas, Brian, ***Acts & Facts Magazine***, March 2017, pg. 20.
4. Hovind, ***Dinosaurs Creation Evolution***, pg. 39.
5. ***Ibid.,*** pg 23.
6. Dillow, Joseph, ***The Waters Above***, Moody Press, Chicago, Ill.: 1981, pg. 120
7. The ***Answers Book***, pg. 119.
8. ***Ibid***., pg. 122.
9. Snelling, Andrew**,** ***AFA Journal***, Is Genesis Really True?, February 2017, pg. 25
10. ***Acts & Facts Magazine***, Institute For Creation Science, January 2017, pg. 9.
11. ***Ibid.***
12. ***Ibid.***
13. ***Ibid.***
14. ***Ibid***.
15. Ham, ***The Answer Book***, pg. 68.
16. ***Acts & Facts Magazine,*** Vernon Cupps, "Radiocarbon Dating Can't Prove an Old Earth" April 2017, pg. 9.
17. Hovind, Ken, ***The Answer Book***, pg. 36.
18. Morris, Henry, pg. 160.
19. Max, Arthur, ***The Jackson Sun***, daily newspaper in Jackson, Tennessee, section A6, December 1, 2011.
20. Rutherford/Plimer ***Austrainar***, Australian geologist professor Corne.
21. Pachauri, Rajendra, UN Chairman of Panel on Climate Change in 2007, when Al Gore won the 2007, Nobel Peace Prize.

22. ***Acts & Facts,*** Vernon Cupps, "Radiocarbon Dating Can't Prove an Old Earth." April 2017, pg. 9.
23. Beisner, Cal, ***One News Now***, Cornwall Alliance for Stewardship for Creation, daily report, June 2, 2013.
24. Ham, Ken, General Editor, ***The New Answers Book 1***, Answers in Genesis, Masters Books, Green Forrest, Arkansas: 2011, pg. 124.
25. Henry M. Morris III, ***The Book of Beginnings,*** Vol. 1, Creation, The Fall and The First Age, Institute of Creation Research, Dallas, Texas: 2012, pg. 34.

Chapter 13: Various Views of Creation Expanded

1. Flamming, Peter James, ***Layman's Library of Christian Doctrine***, Broadman Press, Nashville, Tennessee: 1978, pg. 10.
2. Lisle, Jason, ***Why Genesis Matters***, Christian Doctrine and the Creation Account, The Institute for Creation Research, Dallas, Texas: 2012, pg. 10.
3. Barnes, ***Barnes Notes***, pg. 28.
4. Stanley, Charles, ***Enter His Gates***, A Daily Journal into the Master's Presence, Thomas Nelson Publishers, Nashville, Tennessee: 1998, pg. 109.
5. Johnson, P., ***Defeating Darwinism by Open Minds***, InterVarsity Press, Downers Grove, Illinois: 1997, pg. 15.
6. Morris, ***The Book of Beginnings***, vol. 1, pg. 31.
7. Ham, ***The Genesis Solution***, pg. 45.
8. Ralph O. Muncaster, ***Dismantling Evolution,*** Building the Case For Intelligent Design, Harvest House Publishers, Eugene, Oregon: 2003, pg. 23.
9. Behe, Michael J., ***Darwin's Black Box***, Touchstone Book published by Simon & Schuster, New York, NY: 1996, pg. 78.
10. Boice, James Montgomery, ***An Expositional Commentary***, Genesis, Vol I, Published by Baker Books, Grand Rapids, Michigan: 1982, pg. 42
11. ***Ibid***. pg. 42.

12. Darwin, ***On The Origin of the Species***, pg. 201.
13. ***Ibid.***
14. News Article, ***The Jackson Sun***, What happened to Madeline Murray O'Hair, Mar. 13, 1998.
15. ***The Genesis Solution,*** pg. 14.
16. Morris, Henry M. III, ***The Book of Beginnings***, Institute of Creation Research, Vol. I, Creation, The Fall and the First Age, Dallas, TX: 2012, Vol 1, pg. 8.
17. ***Ibid.***, pg. 46
18. Rhodes, Ron, ***Creation vs Evolution Debate***, Harvest House Publishers, Eugene, Oregon: 2004, pg. 76–77.
19. Kennedy, D. J., ***Origins: Creation or Evolution?,*** Coral Ridge Ministries, Fort Lauderdale, Florida: 1986, pg. 2.
20. Boom, Corrie Ten, ***Each New Day***, A Daily Devotional Study, Fleming H. Revell, Division of Baker Book House, Grand Rapids, Michigan: 1977, pg. 80.
21. Denton, Michael, ***Evolution: A Theory in Crisis***, Alder and Adler, Bethesda, Maryland: 1986, pg. 23
22. Phillips, ***Exploring Genesis***, pg. 44.
23. ***Ibid.***, pg. 49–50.
24. ***Evolution: A Theory in Crisis***, pg. 117.
25. Rhodes, ***Creation vs Evolution Debate***, pg. 58.
26. Boice, James Montgomery, ***An Exposition Commentary***, Genesis, Vol. I. Baker Book House, Grand Rapids, Michigan: 1982, pg. 53.
27. Ham, ***The Genesis Solution***, pg. 67.
28. Martin, Jobe, ***The Evolution of a Creationist***, Biblical Discipleship Publishers, Rockwall, Texas: 1994, pg. 10.
29. Wysong, Randy L., ***The Creation/Evolution Controversy***, Inquiry Press, East Lansing, Michigan: 1976, pg. 63.
30. Boice, ***An Exposition Commentary***, vol. 1, pg. 57.
31. ***Ibid.***
32. Ham, ***The Genesis Solution***, pg. 61–62.
33. Morris, ***The Genesis Record***. pg. 47.

34. Muncaster, ***Dismantling Evolution***, pg. 122.
35. ***Layman Library of Christian Doctrine***, pg. 11.

Chapter 14: The Connection to the Cross

1. Bonhoffer, ***Creation & the Fall***, pg. 93.

CPSIA information can be obtained
at www.ICGtesting.com
Printed in the USA
LVHW011559301020
670212LV00004B/4